D1621718

Data Center Networks
and
Fibre Channel over Ethernet
(FCoE)

Copyright © 2008 by Nuova Systems Inc.
All Right Reserved, including the right of reproduction,
whole or in part, in any form.
Published by Lulu.com
April 2008,
First Edition
Version 1.2

ISBN: 978-1-4357-1424-3

About the author

Silvano Gai, who grew up in a small village near Asti, Italy, has over twenty seven years of experience in computer engineering and computer networks. He is the author of several books and technical publications on computer networking and has written multiple Internet Drafts and RFCs. He is responsible for 30 issued patents and 50 patent applications. His background includes seven years as a full professor of Computer Engineering, tenure track, at Politecnico di Torino, Italy and seven years as a researcher at the CNR (Italian National Council for Scientific Research). For the past eleven years, he has been in Silicon Valley where in the position of Cisco Fellow, he was an architect of the Cisco Catalyst family of network switches and the Cisco MDS family of storage networking switches. Currently, he is a senior fellow with Nuova Systems where he has been active in defining the new Data Center architecture and in making FCoE a reality.

Data Center Networks
and
Fibre Channel over Ethernet
(FCoE)

Silvano Gai

Table of Contents

Preface

This book is the result of the work done by Nuova Systems, in collaboration with Cisco Systems, on the evolution of Ethernet as a Data Center Network in the years 2006 and 2007. The technologies described in this book have been accepted by the industry and, starting in 2008, they will make their way in products and standards.

The book describes them with an educational view: I have tried to place at the readers' disposal updated material compliant with current standards or proposal for future standards. Details will change over time and this book is not intended to be used as a basis for designing products: the appropriate standards must be used.

This book probably contains errors; I would appreciate if you emailed them to: dc_book@ip6.com

I wish to express my thanks to the Nuova Systems and Cisco Systems executive teams who made this book possible, but a particular thank goes to the many engineers and marketing folks with whom I have collaborated in the recent years to develop these technologies.

Finally particular thanks go to Luca Cafiero: he is the father of FCoE.

My gratitude goes to Mario Mazzola who has been my mentor and my friend in all these years.

1.   I/O Consolidation in the Data Center

1.1.   Introduction

Today Ethernet is by far the dominant interconnection network in the Data Center. Born as a shared media standard, Ethernet has evolved over the years and it has become a point-to-point full-duplex network. In the Data Center, it is deployed at speeds of 100 Mbps and 1 Gbps that match reasonably well the current I/O performance of PCI based servers.

A notable exception is represented by the storage traffic that is normally carried over a separate network built according to the Fibre Channel (FC) standard. Most large Data Centers have an installed base of Fibre Channel. These FC networks (also called fabrics) are typically not very large and many separate fabrics are deployed for different group of servers. Most Data Center also duplicate FC fabrics for high availability reasons.

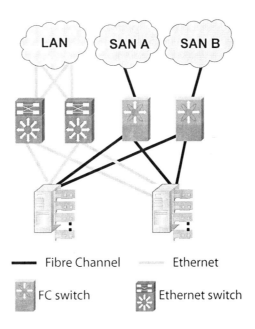

Figure 1: Current Data Center Architecture

In the High Performance Computing (HPC) sector, and for applications that require cluster infrastructures, dedicated and proprietary networks like Myrinet and Quadrix have been deployed. A certain penetration has been achieved by Infiniband (IB), both in the HPC sector and, for specific applications, in the Data Center. IB provides a good support for clusters requiring low latency and high throughput from user memory to user memory.

Figure 1 illustrates a standard Data Center configuration with one Ethernet core and two independent SAN Fabrics for availability reasons (labeled "SAN A" and "SAN B").

1.2. What is I/O Consolidation

I/O consolidation is the ability of a switch or a host adapter to use the same physical infrastructure to carry different types of traffic that typically have very different traffic characteristics and handling requirements.

From the network side this equates in having to install and operate a single network instead of three (Figure 2).

From the hosts and storage arrays side, customers purchase fewer CNAs (Converged Network Adapters) instead of NICs, HBAs and HCAs. This requires a lower number of PCI slots on the servers and it is particularly beneficial in the case of Blade Servers.

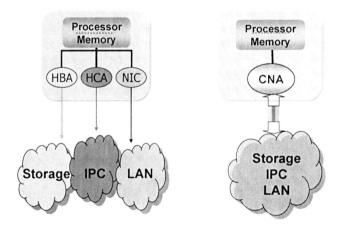

Figure 2: I/O consolidation in the network

The benefits for the customers are:

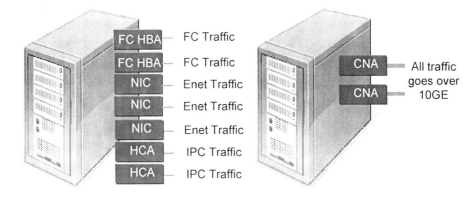

Figure 3: I/O consolidation in the servers

- great reduction, simplification and standardization of cabling;
- absence of gateways that are always a bottleneck and a source of incompatibilities;
- less power and cooling;
- reduced cost.

To be viable, I/O consolidation should maintain the same management paradigm that currently applies to each traffic type.

Figure 3 shows an example where 2 FC HBAs, 3 Ethernet NICs and 2 IB HCAs are replaced by 2 CNAs (Converged Network Adapters).

1.3. Merging the requirements

The biggest challenge of I/O consolidation is to satisfy the requirements of different traffic classes with a single network.

The classical LAN traffic that nowadays is constituted mainly by IPv4 and IPv6 traffic must run on native Ethernet [2]: too much investment has been done in this area and too many applications assume that Ethernet is the underlying network. This traffic is characterized by a large number of flows. Typically these flows were not very sensitive to latency, but this is changing rapidly and latency must be taken into serious consideration. Streaming Traffic is also sensitive to latency jitter.

Storage traffic must follow the Fibre Channel (FC) model. Again, large customers have massive investment in FC infrastructure and management. Storage provisioning often relies on FC services like naming, zoning, etc. In FC losing frames is not an option, since SCSI is extremely sensitive to packet drops. This traffic is characterized by large packet sizes, typically 2KB payload.

IPC (Inter Processor Communication) traffic is characterized by a mix of large and small messages. It is typically latency sensitive (especially the short messages). IPC traffic is used in Clusters, i.e. interconnections of two or more computers. In the Data Center examples of server clustering include:

- Availability clusters (e.g., Symantec/Veritas VCS, MSCS);
- Clustered File Systems;
- Clustered Databases (e.g., Oracle RAC);
- VMware Virtual Infrastructure Services (e.g., VMware VMotion, VMware HA).

Clusters don't care too much about the underlying network, provided that it is cheap, it is high bandwidth, it is low latency and the adapters provide zero-copy mechanisms.

1.4. Why I/O consolidation has not succeeded yet?

There have been previous attempts to implement I/O consolidation. Fibre Channel itself was proposed as an I/O consolidation network, but its poor support for multicast/broadcast traffic never made it credible.

Infiniband has also attempted I/O consolidation with some success in the HPC world. It has not penetrated a larger market due to its lack of compatibility with Ethernet (again, no good multicast/broadcast support), with FC (it uses a storage protocol that is different from FC) and the presence of gateways that are bottlenecks and incompatibility points.

iSCSI has probably been the most significant attempt of I/O consolidation. Up to now it has been limited to the low performance servers, mainly due to the fact that Ethernet had a maximum speed of 1 Gbps. This limitation has been removed by 10GE, but there are concerns that the TCP termination required by iSCSI is onerous at 10Gbps. The real downside is that iSCSI is "SCSI over TCP", it is not "FC over TCP", and therefore it does not preserve the management and deployment model of FC. It still requires gateways and it has a different naming scheme (perhaps a better one, but anyhow different), different zoning, etc.

1.5.   Fundamental technologies

The two technologies that will play a big role in enabling I/O consolidation are PCI-Express and 10GE.

1.5.1. PCI-Express

PCI-Express (PCI-E or PCIe) [1], is a computer expansion card interface format designed to replace PCI, PCI-X and AGP.

It removes one of the limitations that plagued all this I/O consolidation attempts, i.e., the lack of I/O bandwidth in the server buses.

PCIe 1.1 uses full duplex serial links at 2.5 Gbps (2 Gbps at the datalink) and it supports speeds from 2 Gbps (1x) to 32 Gbps (16x). Due to protocol overhead 8x is required to support a 10 GE interface.

PCIe 2.0 (aka PCIe Gen 2) doubles the bandwidth per serial link from 2 Gbit/s to 4 Gbit/s and products have started to ship in 2008.

1.5.2. 10GE

2008 is the year of 10GE. The standard has reached the maturity status and cheap cabling solutions start to be available.

Fiber continues to be used for longer distances, but copper is deployed in the Data Center (lower cost).

Switches and CNAs have standardized their connectivity using the small form-factor pluggable (SFP) transceiver. SFPs are used to interface a network device mother board (i.e, switches, routers or CNAs) to a fiber optic or copper cable. SFP is a popular industry format supported by several component vendors. The standard has expanded to SFP+, which supports data rates up to 10.0 Gbps. Applications of SFP+ include 8 Gbps Fibre Channel and 10GE.

Unfortunately the IEEE standard for Twisted Pair (10GBASE-T) requires an enormous amount of transistors, especially when the distance grows toward 100 meters (328 feet). This translates in a significant power requirement and also in additional delay (Figure 4). Imagine trying to cool a switch line card

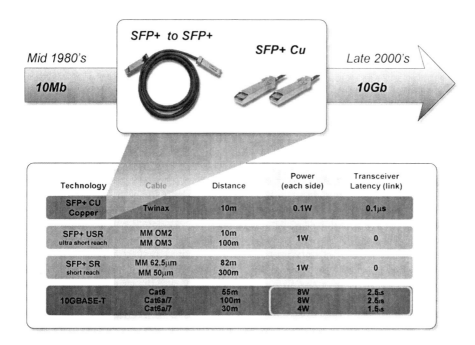

Technology	Cable	Distance	Power (each side)	Transceiver Latency (link)
SFP+ CU Copper	Twinax	10m	0.1W	0.1μs
SFP+ USR ultra short reach	MM OM2 MM OM3	10m 100m	1W	0
SFP+ SR short reach	MM 62.5μm MM 50μm	82m 300m	1W	0
10GBASE-T	Cat6 Cat6a/7 Cat6a/7	55m 100m 30m	8W 8W 4W	2.5.s 2.5.s 1.5.s

Figure 4: Evolution of Ethernet Physical Media

that has 48 10GBASE-T ports on the front-panel, each consuming 4 Watts!

A more practical solution in the Data Center, at the rack level, is to use SFP+ with Copper Twinax cable. The cable is very flexible, approximately 6 mm (1/4 of an inch) in diameter and it uses the SPF+ themselves as the connectors. Cost is limited, power consumption and delay are negligible. It is limited to 10 meters (33 feet) which is sufficient to connect few racks of servers to a common top of the rack switch.

Figure 5 illustrates the advantages of using twinax cable inside a rack or few racks.

The cost of the transmission media is only one of the factors that need to be addressed to manufacture 10GE ports that are really cost competitive. Other factors are the size of the switch buffers, and layer 2 vs. layer 3/4 functionalities.

1.6.   Additional Requirements

1.6.1.   Buffer requirement

Buffer requirement is a complex topic related to propagation delays, higher

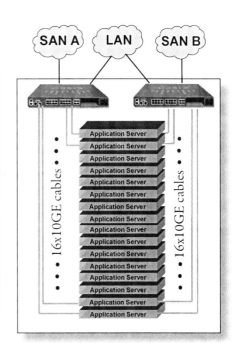

- Low power consumption
- Low cable cost
- Low transceiver latency
- Low error rate (10^{17})
- Thinner cable with higher bend radius

- Easier to manage cabling solution reduces deployment time
- All copper cables are contained within rack

SAN A   LAN   SAN B

16x10GE cables

Application Server
Application Server
Application Server
Application Server
Application Server
Application Server
Application Server
Application Server
Application Server
Application Server
Application Server
Application Server
Application Server
Application Server
Application Server
Application Server

16x10GE cables

Figure 5: Twin-ax Copper Cable

level protocols, congestion control schemes, etc. For the purpose of this discussion it is possible to divide the networks into two classes: "lossless networks" and "lossy networks".

This classification does not consider losses due to transmission errors that, in a controlled environment with limited distances like the Data Center, are rare compared to losses due to congestion.

Fibre Channel and IB are examples of lossless network, i.e. they have a link level signaling mechanism to keep track of buffer availability at the other end of the link. This mechanism allows the sender to send a frame only if a buffer is available in the receiver and therefore the receiver never needs to drop frames. While this seems very attractive at a first glance, a word of caution is in order: lossless networks require to be engineered in simple and limited topologies. In fact congestion at a switch can propagate upstream throughout the network, ultimately affecting flows that are not responsible for the congestion. If circular dependencies exist, the network may experience severe deadlock and/or livelock conditions that can significantly reduce the performance of the network or even completely destroy its functionality. These two phenomena are well known in literature and easy to reproduce in real networks. This should not discourage the potential user, since Data Center networks have simple and well defined topologies.

Historically Ethernet has been a lossy network, since Ethernet switches don't use any mechanism to signal to the sender that they are out of buffers. A few years ago, IEEE 802.3 added a PAUSE frame to Ethernet. This frame can be used to stop the sender for a period of time, but pragmatically this feature has not been successfully deployed. Today it is common practice to drop frame, when an Ethernet switch is congested. Several clever ways of dropping frames and managing queues have been proposed under the general umbrella of AQM (Active Queue Management), but they don't eliminate frame drops and re-quire large buffers to work effectively. The most used AQM scheme is probably RED (Random Early Detection).

Avoiding frame drops is mandatory for carrying native storage traffic over Eth-ernet, since Storage Traffic does not tolerate frame drops. SCSI was designed with the assumption that SCSI transaction are expected to succeed and that failures are so rare that is acceptable to recover slowly from them.

Fibre Channel is the primary protocol used to carry storage traffic and it avoids drop through a link credit mechanism called B2B credits (Buffer-to-Buffer credits). iSCSI is an alternative to Fibre Channels and it solves the same prob-lem by requiring TCP to recover from frame drops, but this implies a heavy use of TCP that has not been widely accepted in the Data Center.

In general, it is possible to say that lossless networks require fewer buffers in the switches than lossy networks and that these buffers may be accommodated on-chip (cheaper and faster), while large buffers require off-chip memory (ex-pensive and slower).

Both behaviors have advantages and disadvantages. Ethernet needs to be ex-tended to support the capability to partition the physical link into multiple logical links (by extending the IEEE 802.1Q Priority concept) and to allow lossless/lossy behavior on a Priority basis.

Finally it should be noted that when buffers are used they increase latency (see Section 1.6.3).

1.6.2. Layer 2 only

A significant part of the cost of a 10 GE inter-switch port is related to function-alities above layer 2, namely IPv4/IPv6 routing, multicast forwarding, various tunneling protocols, MPLS (Multi-Protocol Label Switching), ACLs (Access

Control Lists), and deep packet inspection (layer 4 and above). These features require external components like RAMs, CAMs, or TCAMs that significantly increase the port cost.

Virtualization, Cluster and HPC often require extremely good layer 2 connectivity. Virtual Machines are typically moved inside the same IP subnet (layer 2 domain) often using a layer 2 mechanism like gratuitous ARP. Cluster members exchange large volume of data among themselves and often use protocols that are not IP-based for membership, ping, keep-alive, etc.

A 10 GE solution that is wire-speed, low-latency and completely Ethernet compliant is therefore a good match for the Data Center, even if it does not scale outside the Data Center. Layer 2 domains of 64K to 256K members should satisfy the Data Center requirement for the next few years.

In order to support on the same network multiple independent traffic types, it is crucial to maintain the concept of Virtual LANs and to expand the concept of Priorities (see Section 2.7).

1.6.3. Low Latency

The latency parameter that cluster users really care about is the latency incurred in transferring a buffer from the user memory space of one computer to the user memory space of another computer. The main factors that contribute to the latency are:

1.   The time elapsed between the moment the application posts the data and the moment the first bit starts to flow on the wire. This is basically determined by the zero-copy mechanism and by the capability of the NIC to access the data directly in host memory, even if this is scattered in physical memory. To keep this time low most NICs today use DMA scatter/gather operations to efficiently move frames between the memory and the NIC. This in turn is influenced by the type of protocol offload used: i.e. stateless vs. TOE (TCP Offload Engine).

2.   Serialization delay: this depends only on the link speed. For example, at 10GE the serialization of one Kbyte requires 0.8 microseconds.

3.   Propagation delay: this is very similar in copper and fiber, it is typically 2/3 of the speed of light and can be rounded to 200 meters/microsecond one way, 100 meters/microsecond round-trip delay.

Some people prefer to express it as 5 nanoseconds/meter, and this is also correct. In published latency data, propagation delay is always assumed to be zero. It is also clear that the distance of Data Center Networks must be limited to a few hundreds meters, otherwise this factor becomes dominant and low latency cannot be achieved.

4.   Switch latency depends from the presence or absence of congestion. Under congestion the switch latency is mainly due to the buffering occurring inside the switch and low latency cannot be achieved. In a non-congested situation the latency depends mainly from the switch architecture: store-and-forward vs. cut-through. Today many Ethernet switches are designed with a store-and-forward architecture, since this is a simpler design. Store-and-forward adds several serialization delays inside the switch and therefore the overall latency is negatively impacted. Cut-through switches have a lower latency at the cost of a more complex design.

5.   Same as in point 1, but on the receiving side.

1.6.4.   Native Support for Storage Traffic

The term native support for storage traffic indicates the capability of a network to act as a transport for the SCSI protocol. Figure 6 illustrates possible alternative SCSI transports.

SCSI was designed assuming as the underlying physical layer a short parallel cable, internal to the computer, and therefore extremely reliable. Based on this assumption, SCSI is not efficient in recovering from transmission errors. A frame loss may cause SCSI to time-out and recover in up to one minute.

For this reason, when the need arose to move the storage out of the servers in the storage arrays, the Fibre Channel protocol was chosen as a transport for SCSI. Fibre Channel, through its B2B credit mechanism, guarantees the same frame delivery reliability of the SCSI parallel bus and therefore is a good match for SCSI.

Ethernet does not have a credit mechanism, but it does have a PAUSE frame. A proper implementation of the PAUSE frame achieves results identical to a credit mechanism, in a distance-limited environment like the Data Center.

To support I/O consolidation, i.e. to avoid different classes of traffic to interfere, PAUSE needs to be extended per priority (see Section 2.7).

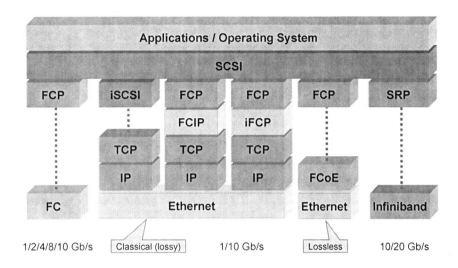

Figure 6: SCSI transports

1.6.5. RDMA support

Cluster applications require two message types:

- Short synchronization messages among cluster nodes with minimum latency.

- Large messages to transfer buffers from one node to another without CPU intervention. This is also referred to as RDMA (Remote Direct Memory Access).

In the latter case the buffer resides in the user memory (in contrast to kernel) of a process and must be transferred in the user memory of another process. User memory is virtual memory and it is therefore scattered in real memory.

The RDMA operation must happen without CPU intervention and therefore the NIC must be able to accept a command to transfer a user buffer, gather it from real memory, implement a reliable transport protocol, and transfer it to the other NIC. The receiving NIC verifies the integrity of the data, signals the successful transfer or the presence of errors, and scatters the data in the destination host memory without CPU intervention.

In the IP world, there is no assumption on the reliability of the underlying network. iWARP (Internet Wide Area RDMA Protocol) is an Internet Engineering Task Force (IETF) update of the RDMA Consortium's RDMA over

TCP standard. iWARP is layered above TCP, which guarantees in order frame delivery. Frames dropped by the underlying network are recovered by TCP through retransmission.

In-order frame delivery is difficult, if not impossible, to satisfy in the Internet, but has been successfully used in networks with a limited scope, like Fibre Channel.

With the advent of Lossless Ethernet, frames are delivered in-order. Dropping happens only as a result of catastrophic events, like transmission errors or topology reconfiguration. The RDMA protocol may therefore be designed with the assumption that frames are normally delivered in order without any frame being lost. Protocols like LLC2, HDLC, LAPB, etc. work well if the frames are delivered in order and if the probability of frame drop is low.

Lossless Ethernet can also be integrated with a congestion control mechanism at layer 2.

Another important point is the support of standard APIs. Among the many proposed, RDS, IB verbs, SDP and MPI seem the more interesting. RDS is used in the database community and MPI is widely adopted in the HPC market.

OFED (Open Fabrics Alliance) is currently developing a unified, open-source software stack for the major RDMA fabrics.

2.   Enabling technologies

Ethernet needs to be enhanced to be a viable solution for I/O consolidation [3], [5], [6]. This section describes the required enhancements.

2.1.   Lossless Ethernet

The term lossless Ethernet has been recently introduced to indicate an implementation of Ethernet bridges (aka switches) that do not lose frame under congestion.

Three questions come to mind immediately ... plus one:

- Can Ethernet be lossless?
- Is a credit scheme required?
- Is lossless better?

... plus ...

- Is anything else required?

2.2.   PAUSE

Historically, there have been three possible causes of frame drop in Ethernet:

- Frame Errors: A frame is received with an incorrect FCS (CRC). These errors are very rare and can only be recovered by a higher level protocol. The probabilities of a frame error in Fibre Channel and in Ethernet are the same at the same speed: Fibre Channel accepts rare frame errors and so does Ethernet. Not a factor in Data Center environments.

- Collisions: A frame cannot be transmitted due to multiple collisions during the transmission attempts. Collisions were possible in Ethernet with shared media. With the advent of Fast Ethernet (100 Mbps) IEEE introduced the full duplex links. Full duplex is the only modality supported in 10 GE and it does not have collisions, since a transmission media is dedicated per each direction and it is always available.

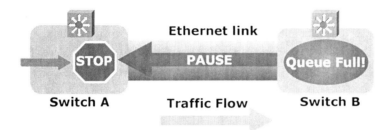

Figure 7: Ethernet PAUSE

- Congestions in switches that cause buffer overflow. These are avoided in Fibre Channel by using buffer to buffer credits (see Section
  2.3). The equivalent mechanism in Ethernet is the PAUSE Frame
  defined in IEEE 802.3 – Annex 31B [2]. The PAUSE operation is
  used to inhibit transmission of data frames for a specified period of
  time when a switch queue (buffer) is full (see Figure 7).

Therefore, a proper implementation of the Ethernet PAUSE may transform a
Ethernet fabric into a lossless fabric.

A PAUSE frame is a standard Ethernet frame, not tagged (the pausing does

PAUSE Frame

01:80:C2:00:00:01
Source Station MAC
Ethertype = 0x8808
Opcode = 0x0001
Pause_Time
Pad 42 bytes
. . .
CRC

Figure 8: PAUSE Frame Format

not apply per VLAN or per Priority, but to the whole link). The format of a PAUSE frame is shown in Figure 8.

The PAUSE frame belongs to the category of MAC Control Frames that are identified by Ethertype = 0x8808. In this category the PAUSE frame is identified by the Opcode = 0x0001 that means PAUSE. The only significant field is the Pause_Time that contains the time the link needs to remain paused, expressed in Pause Quanta (512 bits time). If the link needs to remain paused for a long time, it is customary to refresh the pause by sending periodic PAUSE frames. It is also possible to send a PAUSE frame with Pause_Time = 0 to un-pause the link (i.e., restart transmission, without waiting for the timer to expire).

2.3. Credits vs. PAUSE

Common questions are: "is PAUSE equivalent to credits?", "are credits better?"

Credits are used in Fibre Channel to implement a lossless behavior. On each link, at link initialization the number of buffers is pre-agreed and each end keeps an account of the free buffers. Let's consider the link of Figure 9. Switch A is the sender, and it can transmit a frame only if there is at least one free buffer available in switch B. In the case shown in Figure 9 the Buffer-to-Buffer (B2B) count is 0, and therefore Switch A has to wait until Switch B send a R_RDY (Receiver Ready) to switch A, indicating that a buffer has freed up.

With PAUSE (see Section 2.2, Figure 7) Switch A does not keep track of the buffers available in Switch B and it assumes by default that buffers are available, unless told the contrary by switch B with a PAUSE frame.

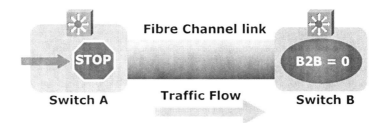

Figure 9: Buffer-to-Buffer Credits

While the technicalities of Credits are different from PAUSE, the observable behavior is the same especially in Data Center environments where the propagation delays are limited.

It should be noticed that for I/O consolidation PFC (see Section 2.7) is superior to FC credits, since FC credits apply to the whole link and not per priority.

2.4. PAUSE propagation

Another question that is often asked is: "How does PAUSE propagate?"

PAUSE, PFC and Credits are all hop-by-hop mechanisms, i.e. they apply to a specific link. They do not automatically propagate to other links in the network.

The goal of these three mechanisms is to suspend the transmission of frames so that the receiver is not forced to drop frames, if it cannot forward them due to congestion.

Let's consider the diagram of Figure 10. Let's assume that switch S3 becomes congested and it issues a PAUSE toward switch S2. S2 suspends transmission and start to build a queue. When the queue exceeds a given threshold, S2 is forced to send a PAUSE to S1 to avoid dropping frames.

Therefore there is not a direct propagation of the PAUSE frame, but it is an indirect mechanism: PAUSE is generated – queue gets full – PAUSE is generated – queue gets full – etc.

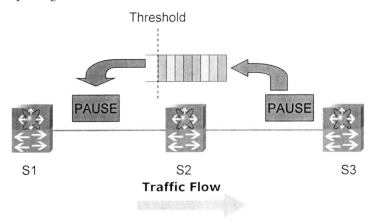

Figure 10: PAUSE Propagation

2.5.   Is lossless better?

This is a complex topic, partially discussed in Section 1.6.1. Lossless has advantages and disadvantages.

On the plus side, frames are never dropped and therefore high level protocols have a reduced amount of work to do. This is particularly important for protocols like SCSI that are not good at error recovery. Lossless is therefore very important for transporting FC over Ethernet. Other application level protocols may take advantage of a lossless behavior, for example NFS (Network File System).

In the case of TCP the fast retransmission takes care of most of the frame losses, but the initial and final frames of the flow are not protected and also the intermediate one may suffer in the presence of severe congestion. Therefore very short TCP flows have been shown to work better on lossless Ethernet.

On the neutral side, TCP relies on losses to adjust its windows and therefore dropping frames is a commonly used technique by AQMs (Active Queue Management) systems like RED (Random Early Detection) to signal congestion to TCP.

On the minus side, lossless can cause congestion spreading and head of line (HOL) blocking and, if circular dependencies exists among buffers, livelock and deadlock. This can be alleviated with techniques like BCN/QCN (see Section 2.8.3). Also care must be taken in configuring the links consistently: a protocol like DCBX (see Section 2.8.1) solves this.

2.6.   Why isn't PAUSE widely deployed?

The main reason is "inconsistent implementations". The IEEE standard defined the basic mechanism, but left the door open to incomplete implementations. This is easy to fix and is being fixed in the new products that are reaching the market.

However, it must not be forgotten that PAUSE applies to the whole link, i.e. it is a single mechanism for all traffic classes. These traffic classes have incompatible requirements (some want lossy, others want lossless) and this may cause "traffic interference"; for example, storage traffic can be paused due to a congestion on IP traffic.

This is clearly undesirable and needs to be fixed.

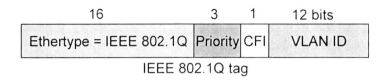

Figure 11: IEEE 802.1Q tag with the Priority field

2.7.   Priority Flow Control (PFC)

Priority Flow Control, aka PPP (Per Priority Pause) [4], is a finer-grain flow control. PFC enables PAUSE functionality per Ethernet priority. IEEE 802.1Q defines a tag (shown in Figure 11), which contains a three bit Priority field, i.e., it can encode eight priorities.

If separate traffic classes are mapped to different priorities, there is no traffic interference; for example, in Figure 12, IPC traffic is mapped to priority six and it is paused, while Storage traffic is mapped to priority three and it is being forwarded and so is IP traffic that is mapped to priority one.

PFC requires a more complex organization in the Data Plane with dedicated resources (e.g., buffers, queues) per priority.

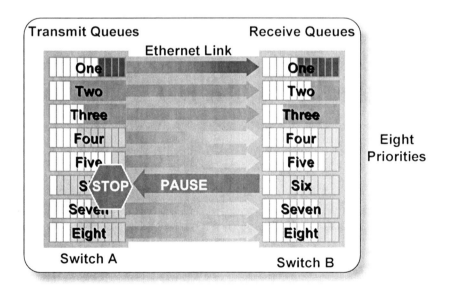

Figure 12: PFC (Priority Flow Control)

PFC is based on a public proposal by Cisco that has a high level of industry support and that is being considered for standardization in IEEE 802.1Qbb.

The PFC Frame format is shown in Figure 13 and it is similar to the PAUSE frame.

The Ethertype = 0x8808 is the same of PAUSE (MAC Control Frame), but the Opcode = 0x0101 is different. There are 8 Time fields, one per priority. To allow for flexible implementation, PFC frames may carry time information for one priority, few priorities or all priorities. This is achieved by having a Class vector with a bit per each priority. For a given priority the bit indicates if the Time field is valid or not.

2.8.   Additional Components

What discussed up to this point are the basic techniques required to implement I/O consolidation. Additional techniques make I/O consolidation deployable on a larger scale. The next paragraphs describe additional components:

*   Discovery Protocol (DCBX);

*   Bandwidth Manager (aka ETS);

*   Congestion Management (aka BCN/QCN).

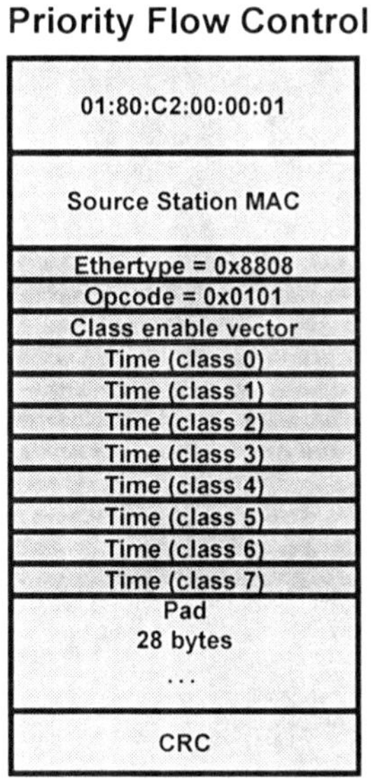

Priority Flow Control

01:80:C2:00:00:01
Source Station MAC
Ethertype = 0x8808
Opcode = 0x0101
Class enable vector
Time (class 0)
Time (class 1)
Time (class 2)
Time (class 3)
Time (class 4)
Time (class 5)
Time (class 6)
Time (class 7)
Pad 28 bytes ...
CRC

Figure 13: PFC Frame Format

2.8.1. DCBX: Data Center Bridging eXchange

DCBX derives its name from an effort in IEEE called DCB (Data Center Bridging), which deals with most of the I/O consolidation techniques described in this document [8]. DCBX is the management protocol of DCB and it is an extension of LLDP (Link Layer Discovery Protocol, IEEE 802.1AB-2005) which is a vendor-neutral Layer 2 protocol that allows a network device to advertise its identity and capabilities on the local network.

DCBX provides hop-by-hop support for:

- Priority Flow Control (PFC);

- Bandwidth Management (ETS);

- Congestion Management (BCN/QCN);

- Applications (FCoE);

- Logical Link Down.

DCBX discovers the capabilities of the two peers at the end of a link, it can check that they are consistent, it can notify the device manager in the case of configuration mismatches, and it can provide basic configuration in the case one of the two peers is not configured.

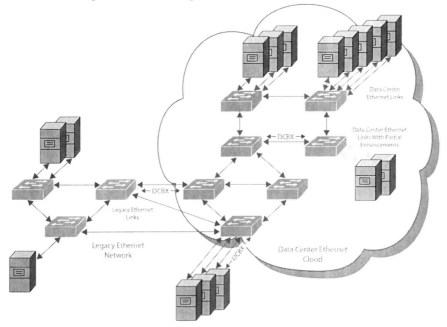

Figure 14: DCBX deployment

Figure 14 shows a deployment scenario for a network that is using DCBX. DCBX-capable links exchange DCB capabilities and conflict alarms are sent to the appropriate management stations. As an example, a boundary is shown indicating which devices support Congestion Management and which do not.

2.8.2. Bandwidth Management

IEEE 802.1Q-2005 defines 8 priorities, but not a simple, effective and consistent scheduling mechanism among them [9]. The scheduling goals are typically Bandwidth, Latency and Jitter control.

Products typically implement some form of Deficit Weighted Round Robin (DWRR), but there is no consistency across implementations, and therefore configuration and interworking is problematic.

IEEE is considering a proposal for a hardware efficient two-level DWRR with strict priority support in IEEE 802.1Qaz ETS (Enhanced Transmission Selection) [10]. Figure 15 shows how priorities are grouped in "Priority Group" with a first level of scheduling and then the Priority Groups are scheduled by a second level scheduler.

With this structure it is possible to assign bandwidth to each "Priority Group", for example: 40% LAN, 40% SAN, and 20% IPC. Inside each Priority Group multiple traffic classes are allowed to share the bandwidth of the group, for example: VoIP and Bulk traffic can share 40% of LAN bandwidth.

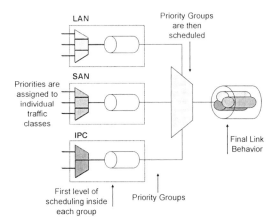

Figure 15: Priority Groups

This architecture allows controlling not only bandwidth, but also latency. Latency is becoming growingly important in particular for IPC applications (see Section 1.6.3).

An example of link bandwidth allocation is shown in Figure 16.

2.8.3. Congestion Management

One of the downside of lossless Ethernet discussed in Section 2.5 is that, in presence of congestion, it tends to create HOL (Head Of Line) blocking, that is undesirable, since it spreads the congestion.

IEEE has activated a standards track project called IEEE 802.1Qau [7] to introduce a layer 2 End-to-End congestion management protocol. The idea behind this effort is to move the congestion from the core to the edges of the network to avoid congestion spreading. At the edge of the network congestion is easier to deal with, since the number of flows is much lower than in the core and therefore the flows that cause congestion can be easily isolated and rate limited.

The algorithms under consideration are BCN (Backward Congestion Notification) and QCN (Quantized Congestion Notification). They are very similar and they act as shown in Figure 17.

The congestion point (CP), i.e., the congested switch, sends messages toward the source of the congestion, i.e., the Reaction Points (RP), to signal a rate reduction through shaping traffic entering the network.

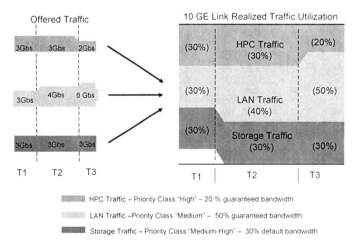

Figure 16: Example of Bandwidth Management

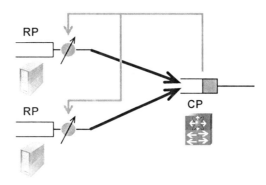

Figure 17: Congestion Point and Reaction Points

A rate limiter is installed as close as possible to the source of the congestion, possibly in the host generating the flow. This alleviates the congestion in the core without causing congestion spreading.

The main difference between this kind of signaling and PAUSE is that PAUSE is hop-by-hop (see Section 2.4), while these messages propagate all the way toward the source of the congestion (see Figure 18).

The rate limiter parameters are dynamically adjusted based on feedback coming from congestion points. This is similar to what TCP does at the transport layer (layer 4 of the ISO model), but it is implemented at layer 2 and therefore it applies to all traffic types, not just TCP. The algorithm used is an AIMD (Additive Increase, Multiplicative Decrease) rate control. In absence of con-

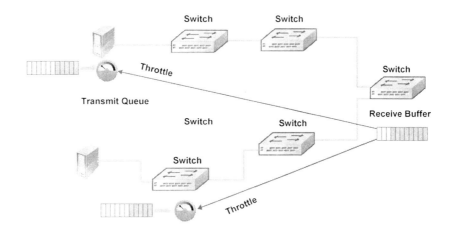

Figure 18: Backward signaling

gestion it increases the bandwidth linearly, but in presence of congestion it decreases it exponentially (gets halved). A similar scheme is implemented for Fibre Channel in the MDS switches and it is called FCC (Fibre Channel Congestion Control).

2.8.4. Layer 2 Multipath

Layer 2 networks forward frames along spanning trees built by the Spanning Tree Protocol (SPT). SPT takes a meshed network and reduces it to a tree by pruning some links. This is clearly undesirable, since it reduces the bisectional bandwidth of the network.

L2 Multipath increases the bandwidth of L2 networks by replacing the STP with IS-IS (an IETF protocol used for IP networks) used at layer 2. IS-IS allows to use all the paths in the network and therefore drastically increases the bisectional bandwidth.

With reference to Figure 19, with layer 2 multipath, both the paths are active and they forward frames, while with STP only one of the two paths is in forwarding state and the other is in hot stand-by.

Multi-path is common in IP networks and it is particularly important when there is limited or no differentiation in speed between access links and backbone links. This is the case for Data Center networks, where initially all links are 10GE.

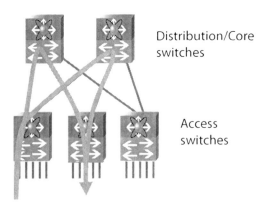

Figure 19: Layer 2 Multipath

Additional advantages of multi-path are:

- multi-path for unicast and multicast frames;
- optimal pair-wise unicast forwarding;
- reduced latency, since less loaded path can be used to forward delay sensitive frames.

Multipath solutions need to provide seamless interoperability with existing protocols, in particular with Spanning Tree.

IETF has a project named TRILL (Transparent Interconnection of Lots of Links) [15] for Layer 2 multi-path and shortest-path frame forwarding in multi-hop IEEE 802.1-compliant Ethernet networks with arbitrary topologies, using IS-IS.

2.8.5. Ethernet Host Virtualizer

The issue of the Spanning Tree Protocol blocking some links in the presence of meshes can also be solved, on a more limited scale, by a technique called Ethernet Host Virtualizer.

With reference to Figure 19, the access switches implement Ethernet Host Virtualizer while the distribution switches continue to run the classical Spanning Tree Protocol.

A switch running Ethernet Host Virtualizer divides the ports in two groups: Server-facing ports and Uplink ports.

Both these types of ports can be a single interface or an Etherchannel.

The switch then associates each server-facing port with an uplink port. This process is called "pinning", and the selected uplink port is called a "Pinned Border Port".

The same server uses always the same pinned border port, unless it fails. In this case, the access switch moves the pinning to another uplink.

The decision of which pinned border port to use for a given server-facing port may be based on manual configuration or decided by the switch based on load. The relationship stays intact until either the server-facing port or the pinned border port loses connectivity.

When this happens, the associated server-facing ports are redistributed to the remaining set of active border ports.

Particular attention must be paid to multicast and broadcast frames to avoid loops and frame duplications. Typically, access switches that implement this feature act as follows:

- they never retransmit a frame received on a border port to another border port;
- they divide the multicast/broadcast traffic into "chunks" (for example, a multicast group may be a chunk);
- they assign a single border port to transmit and receive a broadcast/multicast chunk.

The Ethernet Host Virtualizer is less general than the layer 2 multipath discussed in Section 2.8.3 and only works for access switches that interconnect hosts or storage arrays, not for distribution or core switches. It has the big advantage to be simple to deploy and to plug into an existing core-distribution network without requiring any change.

2.8.6. Delayed Drop

Delayed drop is a means of using PFC to mitigate the effects of short-term traffic bursts while maintaining Packet drop for long-term congestion.

It allows a switch's buffer to virtually extend to the previous hop. This is enabled per priority and it is particularly useful on the lossy priorities to reduce packet drop for transient congestions. It is implemented by asserting PFC on the Priority for a limited period of time.

One of the motivations behind Delay Drop is the limited amount of buffer available in the switches (see Section 1.6.1): being capable of borrowing buffer space from the previous switch may allow absorbing a transient congestion (e.g. an isolated burst) without dropping frames on a lossy priority.

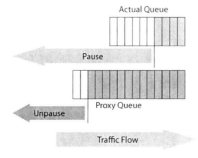

Figure 20: Delay Drop and Proxy Queue

Actual Queue	Proxy Queue
Adds a frame	Adds a frame
Issues a PAUSE	Adds frames at line rate
Drains a frame	Drains a frame
Is Empty	Drains frames at line rate, until empty

Table 1: Delay Drop Actions

When the limited period of time expires, either the burst has been absorbed, (the traffic flows normally again), or frames are dropped like in regular Ethernet.

Figure 20 shows a possible implementation of Delay Drop using a proxy queue to measure the duration of traffic bursts. During normal operation, when packets are added or drained, the proxy queue (that does not really exist, it is just a set of counters) mimics the actual queue. When a burst of traffic is received, the actual queue fill to its high-water marks and issues a PAUSE to stop incoming frames. The proxy queue, that is significatively larger than the actual queue, simulates the continued receipt of frames. When the proxy queue is filled, the PAUSE is released, which in turn causes frame drops. This behavior is summarized in Table 1.

In other words, during short-term congestion both queues drain fast enough that the actual queue releases the PAUSE on its own. During long-term congestion, the proxy queue fills to its high-water mark, and it releases the PAUSE. The actual queue begins to drop packets, and the congestion is managed through higher-level protocols.

2.9.   Nomenclature

All the previously discussed techniques are grouped under different names.

DCB (Data Center Bridging) refers to the standardization activities in IEEE. Projects under consideration in IEEE are:

- Priority Flow Control (see Section 2.7);
- Bandwidth Management (see Section 2.8.2);
- Congestion Management (see Section 2.8.3);

- Configuration (DCBX) (see Section 2.8.1).

DCB specifications submitted to IEEE, but not yet approved, are also called DCB v0 (version zero) and several vendors, in order to introduce products in a timely fashion, claim compatibility with DCB v0.

The terms CEE (Converged Enhanced Ethernet) and DCE (Data Center Ethernet) have also been used to group these techniques under common umbrellas.

Layer 2 Multipath (see Section 2.8.4), Ethernet Host Virtualizer (see Section 2.8.5), Delay Drop (see Section 2.8.6), and Fibre Channel over Ethernet (see Section 3) are complementary, but separated techniques.

3.   FCoE: FC over Ethernet

FCoE (Fibre Channel over Ethernet) is a new standard being developed by the FC-BB-5 working group of INCITS T11 [11], [12], [14]. FCoE is based on the observation that FC is the dominant storage protocol in the Data Center and that any viable I/O consolidation solution for storage must be based on the FC model.

The idea behind FCoE is extremely simple: to implement I/O consolidation by carrying each FC frame inside an Ethernet frame.

Figure 21 shows an example of I/O consolidation with FCoE in which the only connectivity needed on the server is Ethernet, while separate backbones can still be maintained for LAN and SAN.

FC traffic shares Ethernet links with other traffics, as shown in Figure 22.

From a Fibre Channel standpoint, it is FC connectivity over a new type of cable called an "Ethernet cloud", from an Ethernet standpoint it is yet another ULP (Upper Layer Protocol) to be transported in parallel to IPv4, IPv6, IPX, etc. (see Figure 23).

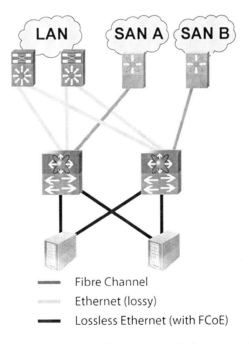

Figure 21: Example of I/O Consolidation with FCoE

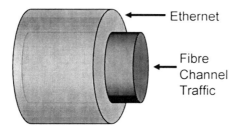

Figure 22: FCoE link sharing

To better understand FCoE, let's have a look at the FCoE protocol stack of Figure 24.

SCSI continues to be mapped over FC as in classical Fibre Channel and there is an encapsulation layer (called FCoE) to encapsulate FC over Ethernet.

The encapsulation is a frame-by-frame encapsulation and therefore the FCoE layer is completely stateless and it does not require fragmentation and reassembly.

FCoE poses some requirements on the underlying Ethernet network, the most important one being lossless. Lossless can be simply achieved by the PAUSE frame (see Section 2.2). More realistically, in an I/O consolidation environment, PFC is used (see Section 2.7) and additional techniques like DCBX (see Section 2.8.1) facilitate deployment. FCoE requires also support of Jumbo frames, since FC frames are not fragmented (see Section 3.3).

3.1.   FCoE Benefits

FCoE benefits are similar to those of other I/O consolidation solutions, i.e., fewer cables, both block I/O & Ethernet traffic co-exist on same cable, fewer adapters needed, and overall less power.

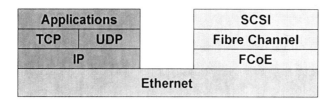

Figure 23: ULPs over Ethernet

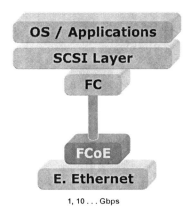

Figure 24: FCoE Protocol Stack

To these generic benefits FCoE adds additional advantages of being completely part of the Fibre Channel architecture, i.e., seamless integration with existing FC SANs, reuse of existing FC SAN tools and management constructs.

Another important advantage is that FCoE requires no gateway. In fact the encapsulation/de-encapsulation functions simply add or remove an Ethernet envelop around a FC frame: the FC frame is untouched and the operation is completely stateless.

From a Storage Administrator perspective, zoning is a basic provisioning function that is used to give Hosts access to Storage. FCoE switches continue to offer an unmodified zoning function ensuring that storage allocation and security mechanisms are unaffected.

The same consideration applies to other Fibre Channel services, for example:

- The dNS (domain Name Service), i.e. the Name Server Database is a FC distributed database that applies as well to FCoE-based SANs;

- RSCN (Registered State Change Notification) functionality is fully supported on FCoE and FCoE adapters receive RSCN exactly like FC HBAs.

- FSPF (Fibre Channel Shortest Path First) is used to compute FC forwarding in a mixed environment FC/FCoE.

A FCoE-connected server is just a SCSI initiator over FC, exactly as the server were connected over native FC (see Section 3.10). The same applies to FCoE-connected Storage Arrays: they are just SCSI targets over FC. The management tools that customers use to manage and maintain their SANs today can be used in a FCoE environment.

Services utilizing storage virtualization or server virtualization continue to work with FCoE, since everything from Fibre Channel up remains intact.

3.2.   FCoE: Protocol Organization

FCoE is really two different protocols:

- FCoE itself is the data plane protocol. It is used to carry most of the FC frames and all the SCSI traffic. This is data intensive and typically it is switched in hardware.

- FIP (FCoE Initialization Protocol) is the control plane protocol. It is used to discover the FC entities connected to an Ethernet cloud and to login to and logout from the FC fabric. This is not a data intensive protocol and it is typically implemented in software on the switch supervisor processor.

The two protocols have two different Ethertypes.

3.3.   FCoE Data Plane

As previously stated, the FCoE Data Plane requires Lossless Ethernet to match the lossless behavior guaranteed in Fibre Channel by buffer-to-buffer credits.

The FCoE encapsulation is shown in Figure 25.

Starting from the inside out, there is the FC Payload that can be up to 2 KB, hence the requirement to support Ethernet Jumbo frame. The FC Payload is wrapped in the FC frame that contains an unmodified FC header and the original CRC. Next is the FCoE header and trailer that mainly contain the

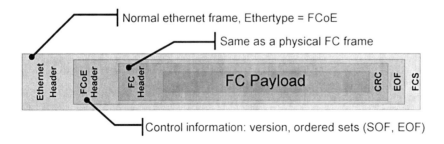

Figure 25: FCoE Encapsulation

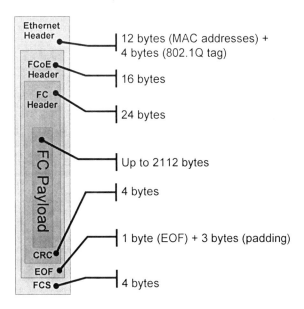

Figure 26: FCoE Frame Size

encoded FC start of frame and end of frame (in native FC these are ordered sets that contain code violation and therefore they need to be re-encoded). Finally the Ethernet header which contains Ethertype = FCoE and the Ethernet trailer that contains the FCS (Frame Control Sequence).

Figure 26 shows the field sizes: the maximum size of a FCoE frame is 2180 bytes. To provide some margin for growth, FCoE requires that the Ethernet infrastructure supports jumbo frames up to 2.5 KB (baby jumbo frames).

Figure 27 shows a detailed view of an Ethernet frame that includes an FCoE frame.

The first 48-bits in the frame are used to specify the Destination MAC address and the next 48-bits specify the Source MAC Addresses. The 32-bit IEEE 802.1Q Tag provides the same function as it does for Virtual LANs, allowing multiple virtual networks across a single physical infrastructure. It also includes the Priority field that must be present to be able to deploy PFC (see Section 2.7).

FCoE has its own Ethertype as designated by the next 16 bits, followed by the 4-bit FCoE version field. The next 100-bits are reserved and have been inserted so that, even in the presence of minimum size FC frames, the Ether-

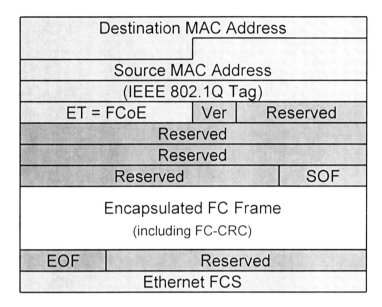

Figure 27: FCoE Frame Format

net payload is greater or equal to 46 bytes, thus avoiding the need to pad the Ethernet Payload. In fact, FCoE frames are never padded.

The 8-bit Start of Frame (SOF) follows and it is encoded as in FCIP (see RFC 3643). The next field is the actual FC frame including the FC CRC, followed by the 8-bit End of Frame (EOF) delimiter, again encoded as in FCIP (see RFC 3643).

The EOF is followed by 24 reserved bits and the frame ends with the final 32-bits dedicated to the Ethernet FCS.

The FC header is maintained unmodified so that when a traditional FC SAN is connected to an FCoE-capable switch the frame is easily encapsulated/ de-encapsulated. This capability enables FCoE to integrate with existing FC SANs without the need of a gateway.

3.4.   FCoE Models

The FCoE entities are modeled in a way similar to the FCIP entities described in FC-BB-4, since FCoE is part of the evolution of FC-BB-4 called FC-BB-5. They also derive the terminology from the same standard.

The first element being defined is the ENode (FCoE Node): a Fibre Channel

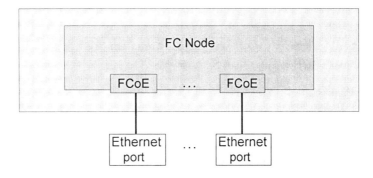

Figure 28: ENode: Simplified Model

HBA implemented within an Ethernet NIC, commercially known as a CNA (Converged Network Adapter).

The simplified model of an Enode with two ports is shown in Figure 28.

The Enode contains a single FC node with multiple "FCoE entities", one per each Ethernet port. Each FCoE Entity has an Ethernet MAC address that is used as a source or destination address when FCoE frames are transmitted through an Ethernet fabric (further details in Section 3.6).

The other element being defined is the "FCoE switch" shown in Figure 29.

An FCoE switch contains both an Ethernet switch function (Ethernet Bridge) and a Fibre Channel switch function or FCF (Fibre Channel Forwarder). The FCoE entity inside the FCF encapsulates FC frames into FCoE Frames and de-encapsulates FCoE frames back to FC frames. FC traffic flows into one end

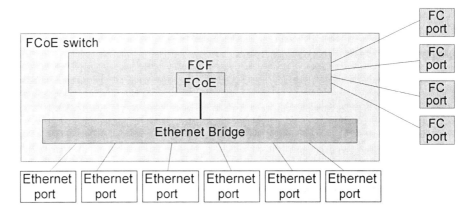

Figure 29: FCoE switch: Simplified Model

of the FCoE entity and FCoE traffic flows out of the other end. As in the case of the Enode the FCoE Entity has an Ethernet MAC address that is used as a source or destination address when FCoE frames are transmitted through an Ethernet fabric (further details in Section 3.6).

For example, in Figure 29, an FC frame enters from the FC ports on the right into the FCF where it is forwarded according to FC rules. If the FC destination is reachable through FCoE, the FC frame is passed to the FCoE entity that encapsulates the Fibre Channel frame into an FCoE frame. The FCoE frame is sent to the Ethernet Bridge where it is distributed to the appropriate Ethernet port on the bottom of the diagram.

The reverse is also true: FCoE traffic flows from the Ethernet ports, through the Ethernet Bridge until it is de-encapsulated inside of the FCoE Entity. The FCF performs a regular FC forwarding operation and may send the frame on one of the FC ports.

This is just one example of how a switch can be FCoE enabled. Different vendors may choose to implement FCoE in different ways. The main difference is the presence (or absence) of one or more bridges inside the FCoE switch.

It is also clear that this architecture fits very well the general architecture of a multiprotocol router and therefore it is easy to foresee that the FCF functionality may be added to a multiprotocol router (see Section 3.13.1).

Figure 28 and Figure 29 are simplified models with informal terminology. Being a bit more formal and complete, the FC-BB-5 standard leverages few concepts already introduced for FCIP and, in particular, the concept of Virtual port.

A Virtual port is a FC port that is implemented on a network that is not native FC. In the case of FCIP the network is IP, in the case of FCoE the network is Ethernet. Therefore FC-BB-5 defines:

- VN_Port (Virtual N_Port): an N_Port over an Ethernet link;
- VF_Port (Virtual F_Port): an F_Port over an Ethernet link;
- VE_Port (Virtual E_Port): an E_Port over an Ethernet link.

It also defines FCoE_LEP (FCoE Link Endpoint) as the data forwarding component that handles FC frame encapsulation/de-encapsulation, and transmission/reception of FCoE frames.

With these definitions in mind it is possible to analyze the detailed model of an Enode shown in Figure 30, where the wide brackets indicate optional components.

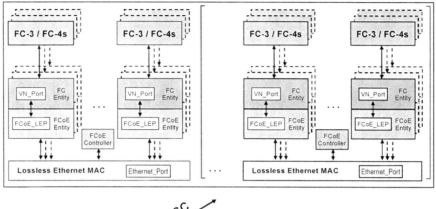

Figure 30: ENode: Complete Model

The Enode consist of one or more Lossless Ethernet MAC, one per physical Ethernet port present on the Enode (CNA).

On each lossless Ethernet MAC there is an entity called FCoE controller that implements the FIP protocol and instantiates the VN_Ports.

VN_Ports may be instantiated as a result of a FLOGI or FDISC operation. A FCoE Enode may decide to FLOGI in one or more FCFs reachable through the Ethernet cloud. Additional software present on the host (e.g. virtualization software) may decide to perform additional FDISCs, for example one per Virtual Machine. Since FCoE supports multiple FLOGI on the same Ethernet port, the FDISCs may be replaced with FLOGI and any arbitrary combination of the two can be used. In the model the VN_Port created left to right represents the result of FLOGI, while the dashed VN_Port behind the main "FLOGI VN_Port" represents the result of FDISC.

The VN_Port remains a strictly FC concept, its adaptation to FCoE is done by the FCoE entity that contains the FCoE_LEP. The FCoE entity is the entity that is addressable at the MAC layer.

The detailed model of an FCoE switch (shown in Figure 31) is very similar. The main differences are the absence of the FC-3/FC-4 layers (higher layers in the FC stack) that are replaced by a regular FC Switching Element (which in the simplified model was called FCF). The FC ports that in the simplified model were on the right are now on top and there is the clear indication that they can act as E_Ports or F_Ports.

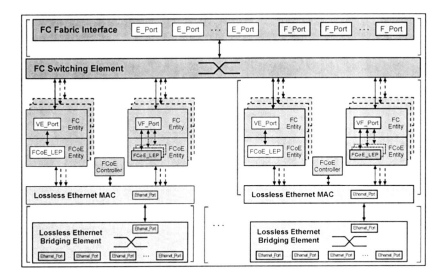

Figure 31: FCoE Switch: Complete Model

The additional difference is that below each Lossless Ethernet MAC there is an optional Lossless Ethernet Bridging Element. This is not strictly needed for FCoE, but it is of paramount importance to provide a global I/O consolidation solution.

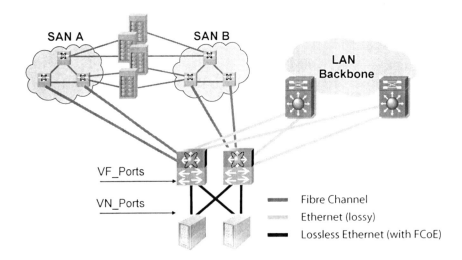

Figure 32: FCoE: Initial deployment

3.5. FCoE: topologies

FCoE can be deployed in a variety of topologies, depending on the business needs of a company and the products chosen. Figure 32 depicts a simple FCoE topology where I/O consolidation is achieved at the access layer of the network.

In this scenario FCoE is used to consolidate traffic from discrete servers to FCoE enabled switches. The servers contain the FCoE Enodes (CNAs) that implement the VN_Ports, while the FCoE switches implement the VF_Ports.

The FCoE switches pass FC traffic to the attached FC SANs and Ethernet traffic to the attached LAN. This deployment model provides the greatest value for a customer that has a large installed base of LAN and SAN environments, since it allows a phased approach to I/O consolidation.

Figure 33 shows a similar approach for blade server deployment, where the only switching element that needs to be present inside the blade server is a lossless Ethernet Bridge. The CNAs on the Server Blades implement the VN_Ports, while the FCoE switches implement the VF_Ports. The Lossless Bridges inside the Blade Server do not need to be FCoE-aware, since they act purely at the Ethernet layer.

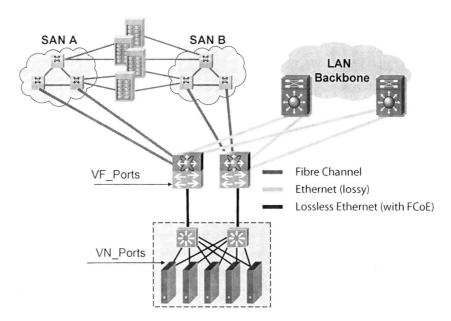

Figure 33: FCoE: Adding Blade Servers

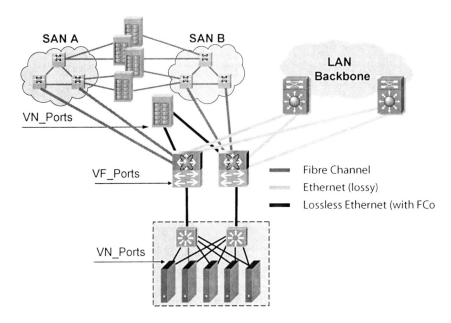

Figure 34: FCoE: Adding Native FCoE Storage

Figure 34 adds FCoE directly connected Storage Arrays. From a FCoE perspective there is no real difference between a host and a storage array. The concept of Initiator and Target are SCSI concepts and not FC concepts and therefore they don't show up in FCoE.

For FCoE a Storage array is just another device that accesses the FCoE network using a CNA. FCoE does not care that the CNA acts as a Target or as an Initiator. From an FC perspective an FCoE storage array is a VN_Port that accesses a VF_Port, exactly like a host. This is identical to the native FC model.

 Another possible topology is shown in Figure 35, where the FC E_Ports between the FCoE switches and the FC switches have been replaced by FCoE VE_Ports. This also shows that it is not only possible for Ethernet switches that implement FCoE to have FC ports, but also for FC switches that implement FCoE to have Ethernet ports.

There are many other FCoE deployments that are possible, for example to extend the reach of FCoE into the aggregation and core layers of the Data Center network. Yet another example is an end-to-end deployment of an entirely FCoE network. All of these topologies are supported by the proposed T11 FC-BB-5 standard.

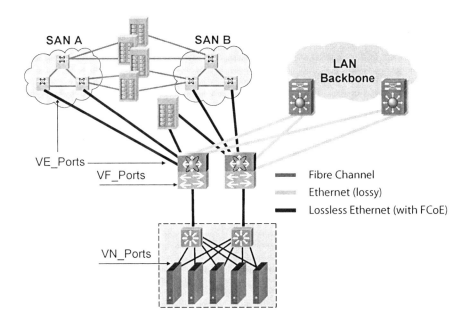

Figure 35: FCoE: Adding VE_ports

3.6.   FCoE addressing

FCoE frames have two types of addresses: MAC addresses and FC addresses.

They serve two different purposes:

- MAC addresses are used as hop-by-hop addresses from the VN_ Port to the VF_Port, or between two VE_Ports, or from a VF_Port to a VN_Port;

- FC addresses (aka FC_IDs or N_Port_IDs) are used as end-to-end addresses from the (V)N_Port of the host to the (V)N_Port of the storage array or vice versa (the V is between parenthesis, since the N_Port may be on FCoE or on FC).

This is consistent with the IP model in which IP addresses are end-to-end and MAC addresses change hop-by-hop.

FC addresses are assigned only to (V)N_Ports, since they are end-to-end, and this is the reason why they are also called N_Port_IDs. (V)E_Ports are never explicitly addressed at the FC layer and therefore they don't have FC addresses.

MAC addresses, being hop-by-hop, are assigned to VN_Ports, VF_Ports and

VE_Ports, since a hop may be the connection between a host and a switch, or between two switches.

Traditional Fibre Channel fabric switches maintain forwarding tables based on FC_IDs. FC switches use these forwarding tables to select the best link available for a frame so that the frame reaches its destination port. Fibre Channel links are typically point-to-point and do not need an address at the link layer.

Ethernet networks are different because Ethernet switches (which are not explicitly addressed) create a "cloud", i.e. a multi-access network. This requires FCoE to rely on Ethernet MAC addresses to direct a frame to its correct Ethernet destination.

Figure 36 shows an example of a frame traveling left to right.

On the left is a storage array attached to a Fibre Channel switch labeled FC Domain 7. This storage array is in a traditional SAN and stores information for a host on an FCoE enabled Fabric. The host has both a FC_ID 1.1.1 and a MAC address C.

The Fibre Channel N_Port on the storage array sends out the FC frame, which includes the Destination FC_ID (D_ID = 1.1.1) and the Source FC_ID (S_ID = 7.1.1) in the FC header (for simplicity, only the header information is displayed in the figure).

The Fibre Channel switch with Domain ID = 7 receives the frame. Since the destination ID (D_ID) in not in its FC domain (it is in domain 1, not in domain 7), the switch looks up the destination domain ID in its forwarding table and transmits the frame on the port associated with the shortest path, as determined by the Fabric Shortest Path First (FSPF) algorithm.

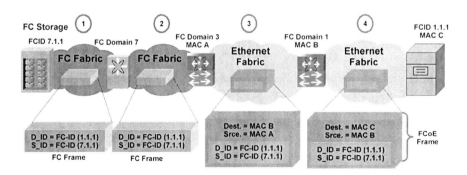

Figure 36: Example of Addressing

The switch with the FC Domain ID = 3 receives the frame and determines that the destination ID (D_ID) is not in Domain 3 and performs the lookup process. However, in this case the FC frame is transmitted across an FCoE enabled Ethernet fabric. This requires the FC frame to be encapsulated into an Ethernet frame by the switch FCoE entity and then transmitted on the port associated with the shortest path.

While the original FC source and destination ID's are maintained in the encapsulated FC frame, the FCoE entity inserts Ethernet and FCoE headers. In the Ethernet header it populates the destination and source MAC addresses. In the example, the destination is MAC address is B (the MAC address of the FCoE entity in the receiving switch) and source MAC address is A (the MAC address of the FCoE entity in the transmitting switch).

When the FCoE frame arrives at the FCoE entity with MAC address B, the frame is de-encapsulated and the switch determines that the FC frame destination is within its domain (Domain 1). The FC frame is re-encapsulated with the new destination MAC address C (which corresponds to the FC D_ID 1.1.1) and the source MAC address is set to B. Then the frame is transmitted out of the appropriate port to the FCoE host with destination MAC address C.

When the frame is received by the CNA (Converged Network Adapter) with MAC address C, the FCoE frame is de-encapsulated and the FC frame accepted by the host with FC_ID 1.1.1.

This example demonstrates how traditional FC addresses map to FCoE MAC addresses. Topologies vary depending on a customer's implementation requirements and the FCoE-capable products deployed, but the addressing schemes remain the same.

3.7. FCoE forwarding

To understand how forwarding works in presence of the Spanning Tree Protocol (SPT) and FSPF, let's consider Figure 37.

STP runs on all the Ethernet clouds and prunes them to trees. In the example we have two spanning trees (STP#1 and STP#2). FSPF runs over the pruned trees and provides end-to-end forwarding at the FC layer. In the previous example there are no meshes and therefore the two STPs do not prune any link and FSPS has a single forwarding path.

Figure 38 shows the same topology with an additional link that has caused the

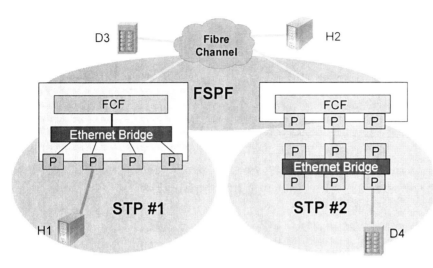

Figure 37: STP/FSPF interaction (1)

two STPs to merge into a single one. The STP does not prune any link, since there are no meshes at the Ethernet layer, but FSPF now sees two alternative paths, a native FC one and a FCoE one: if the metrics are equal, FSPF can balance the traffic between them.

If an additional Ethernet link is added at the bottom (see Figure 39), Ethernet now has a mesh.

STP prunes a link by blocking a port (marked with a "X"). Traffic does not

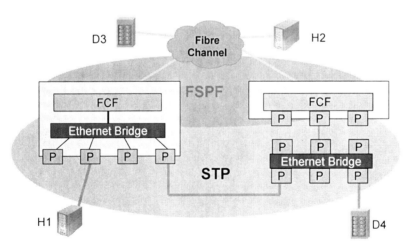

Figure 38: STP/FSPF interaction (2)

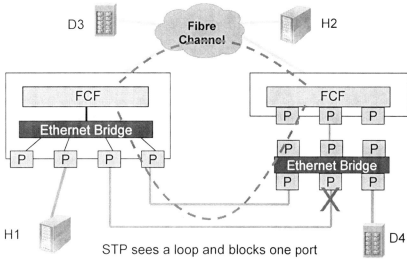

Figure 39: STP/FSPF interaction (3)

flow over the blocked port, unless the parallel Ethernet link fails. FSPF continues to see two alternative paths only (it does not see the blocked link).

Figure 40 shows an alternative approach in which the two parallel links have been grouped using Ethernet Link Aggregation (aka Etherchannel), they are

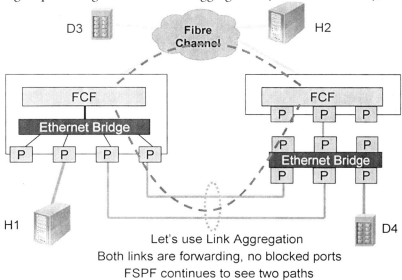

Figure 40: STP/FSPF interaction (4)

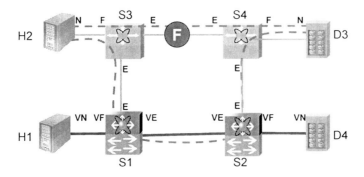

Figure 41: Rerouting in the presence of a fault

both forwarding, since they are seen by STP as a single link. FSPF continues
to see two alternative paths only.

The last example (Figure 41) shows a mixed FC/FCoE topology. Normally the
shortest path from H2 to D3 is the straight FC path through S3 and S4, but
since the link between S3 and S4 has failed, FSPF reroutes the traffic from H2
to D3 through the path H2 – S3 – S1 – S2 – S4 – D3, which is a mixed path
with some native FC links and some FCoE links.

3.8. FPMAs and SPMAs

Let's now discuss how MAC addresses are assigned.

The MAC addresses on the switches, i.e. the MAC addresses associated to the
VE_Ports and VF_Ports, are derived from the switch pool. These are Universal MAC addresses that the switch manufacturer has bought from IEEE and
burned in a ROM (Read Only Memory) inside the switch. They are world-
wide unique.

VN_Ports may use two types of MAC addresses: SPMAs (Server Provided
MAC Addresses) or FPMAs (Fabric Provided MAC Addresses).

FPMAs (aka Mapped MAC Addresses) follow the FC model in which the
switch (i.e. the fabric) assigns the FC_IDs. FPMAs are MAC addresses assigned by the switch during the FC login process (FLOGI/FDISC). They
are not unique world-wide, but the FC fabric guarantees that they are unique
within the fabric, as it does for FC_IDs.

SPMAs are based on the idea that MAC addresses belong to the hosts and
therefore each host decides which MAC address to use. The uniqueness is del-

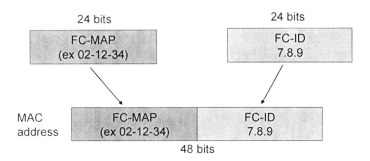

Figure 42: The Mapped MAC addresses

egated to host management software since, with the introduction of virtualization, MAC addresses are no longer unique (for example, see [16]).

The MAC address scheme used by each VN_Port and the MAC Address value are negotiated in FIP (see Section 3.9).

While two schemes are allowed by the standard, initial deployment uses FPMA only.

FPMA addresses are constructed by concatenating a prefix called FC-MAP (Fibre Channel MAC Address Prefix) with the VN_Port FC_ID, as shown in Figure 42.

With FPMA each FC_ID has its own MAC address that is algorithmically derived and therefore there is no need to store it. Since the FC_ID is assigned by the fabric during the FLOGI/FDISC process and the FC-MAP is assigned by the fabric in the FIP protocol, the MAC address is implicitly assigned by the fabric. This is consistent with the FC model.

The FC-MAP is pragmatically an OUI (Organization Unique Identifier) with U/L bit set to 1 to indicate that it is a Local address, not world-wide unique.

Since many customers deploy physically separated SAN fabrics, they may assign different values of FC-MAP to each fabric, thus having an additional assurance that there are no addressing conflicts, if two fabrics are accidentally merged together.

Figure 43 clarifies which addresses are used by FIP and FCoE. The term MAC(Enode) indicates a universal MAC address associated with the Enode, that is used by FIP; the term FCF-MAC indicates the universal MAC address associated with the FCoE switch; All-FCF is a multicast MAC address that groups all the FCoE switches.

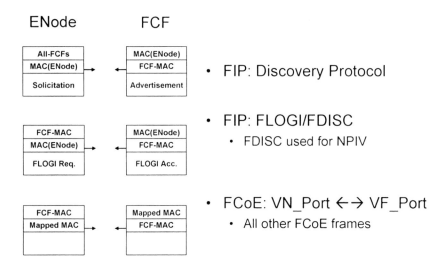

ENode   FCF

All-FCFs		MAC(ENode)
MAC(ENode) →	←	FCF-MAC
Solicitation		Advertisement

- FIP: Discovery Protocol

FCF-MAC		MAC(ENode)
MAC(ENode) →	←	FCF-MAC
FLOGI Req.		FLOGI Acc.

- FIP: FLOGI/FDISC
  - FDISC used for NPIV

FCF-MAC		Mapped MAC
Mapped MAC →	←	FCF-MAC

- FCoE: VN_Port ←→ VF_Port
  - All other FCoE frames

Figure 43: FCoE: Mapped MAC Addresses Usage

An additional advantage of FPMAs is that the MAC address pair encodes also the type of communication (from which port type to which port type) and this adds robustness in the case of erroneous configurations (see Table 2).

VE_Port to VE_Port	DA = Not-Mapped
	SA = Not-Mapped
VF_Port to VN_Port	DA = Mapped
	SA = Not-Mapped
VN_Port to VF_Port	DA = Not-Mapped
	SA = Mapped
VN_Port to VN_Port	DA = Mapped
	SA = Mapped

Table 2: FCoE: Mapped MAC Addresses Usage

3.9.   FIP: FCoE Initialization Protocol

The FIP (FCoE Initialization Protocol) is a control protocol to discover FCoE capable devices connected to an Ethernet network and to negotiate capabilities, including the MAC addresses to be used.

Figure 44 shows a possible flow ladder for FCoE. It starts with the Discovery Phase of FIP in which the CNAs multicast solicitation messages and the FCoE Switches reply with advertisement messages. This interaction takes care of both discovering the FCoE entities and negotiating the capabilities.

FIP next step is the login in the fabric accomplished by the FC command FLOGI/FDISC. The acceptance of these commands by the FCOE switch allows starting the exchange of data traffic.

This is accomplished in the FCoE protocol that carries mostly the SCSI traffic, but also FC control frames like PLOGI, PRLI, RSCN, DNS, etc.

Figure 45 shows the FIP frame format that is characterized by its own Etherype = FIP.

The message body contains a FIP operation illustrated in Figure 46.

The FIP operation code together with the FIP Subcode identifies the particular FIP frame (e.g. Solicitation, Advertisement). Each frame is a set of TLVs (Type-Length-Value) and the Descriptor Length List specifies the length of the TLV list. The Flags are currently unused. The capability bits indicate the

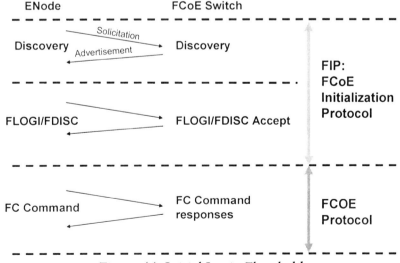

Figure 44: Initial Login Flow ladder

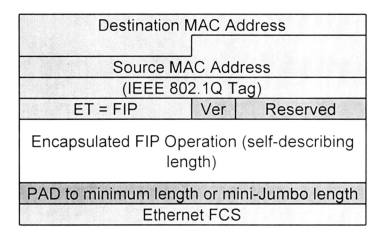

Figure 45: FIP Frame: Contains FIP Operation

type of addressing scheme used (see Section 3.6): in particular the FP bit indicates the support of FPMAs (Fabric Provided MAC Addresses) and the SP bit indicates the support for SPMAs (Server Provided MAC Addresses).

The S bit indicates if the frame is a reply to a solicitation message. The F bit indicates if the frame is originated by a FCF (FCoE switch).

Figure 47 shows the descriptors used by the FIP protocol. As mentioned early each descriptor is encoded with the TLV (Type-Length-Value) syntax. Several descriptors are combined together to build the different messages of the FIP protocol.

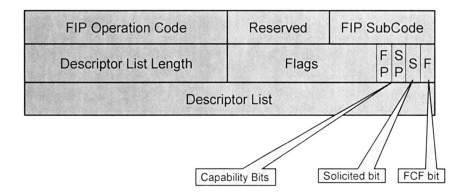

Figure 46: FIP operation format

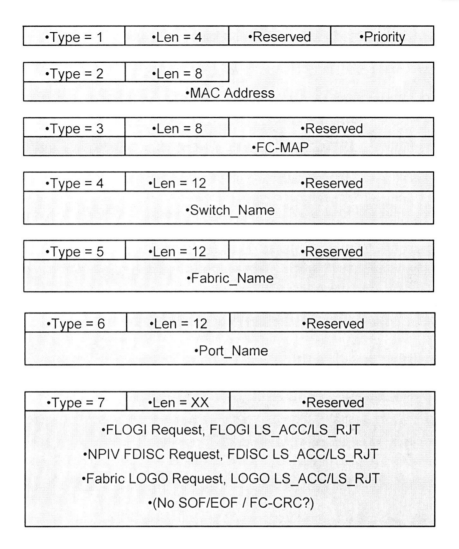

Figure 47: FIP Descriptors

Figure 48 shows the Discovery Solicitation phase and, in particular, the solicitation performed by the host H2.

H2 sends a solicitation message shown in Figure 49 to the multicast MAC address All-FCF-MACs. This message reaches both the FCoE switches (FCF-A and FCF-B). They individually reply as shown later.

 In the solicitation message F is equal to zero, since the message is not generated by a FCoE switch and the FP and SP bits indicate the MAC address scheme supported by the CNA (at least one of the two bits must be set).

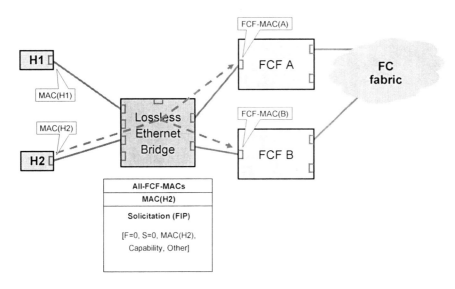

Figure 48: FIP: Multicast Solicitation from H2

This solicitation message is used by the FCoE switch to learn about the Enodes (CNAs), while the one shown in Figure 50 is used by the FCoE switches to discover other FCoE switches directly reachable over an Ethernet cloud.

The purpose of a solicitation done by an FCoE switch (F bit equal 1) is to discover other FCoE switches and to create VE_Port with them. This solicitation indicates which FC-MAP is used by the FCF (i.e. to which fabric it belongs): a receiving FCF ignores this solicitation if it is coming from a different fabric,

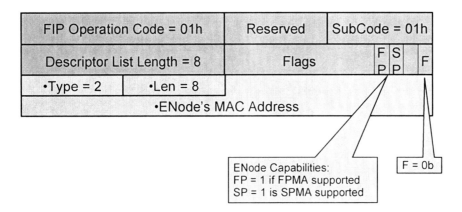

Figure 49: Solicitation from ENode

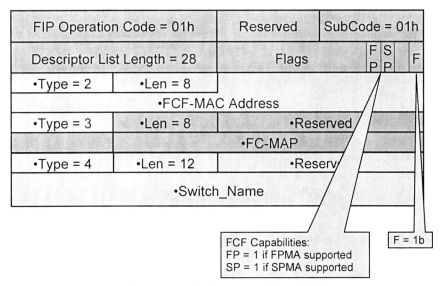

Figure 50: Solicitation from FCF

i.e. if the FC-MAP does not match. Finally, the switch name is also carried in the solicitation.

Let's now consider how an FCF replies to an Enode (CNA) solicitation with

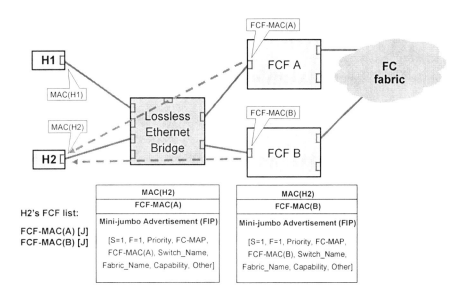

Figure 51: Unicast Advertisements from A and B

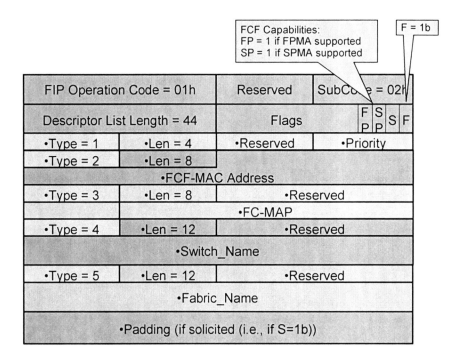

Figure 52: Advertisement

an advertisement. Figure 51 shows the replies of FCF-A and FCF-B to the solicitation from H2.

The replies from the FCF are called "Advertisements" and are sent in unicast. Therefore, in the previous example, they reach only H2 but not H1. In other words, the FCF generates specific advertisement for each Enode. The FCF may also decide to not reply to an Enode solicitation, for example because the capabilities of the Enode and the FCF do not match, or because it is forbidden to do so by security policies.

Figure 52 shows the structure on a FIP advertisement message. The F bit is set to one to indicate that the message comes from an FCF. At least one of the two address capability bits (FP and SP) is set. A priority field indicates the priority of this FCF compared to other FCFs connected to the same fabric. Additional TLVs carry the FCF MAC address, the FC-MAP, the Switch name and the Fabric Name. All these fields are precious for the Enode to perform consistency checks and be sure to login in the right FCF.

Let's now discuss how the FP and SP bits are set. There are basically two cases:

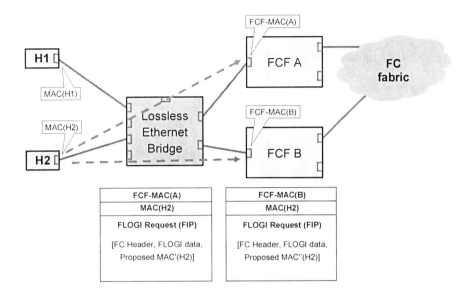

Figure 53: FLOGI Request

- The Enode supports only one addressing scheme, i.e. either SPMA or FPMA. It indicates this in the solicitation and the FCF replies to the solicitation only if it supports the same addressing scheme. Only one bit (FP or SP) is set in both the solicitation and the advertisement, and the associated addressing scheme is used.

- The Enode supports both addressing schemes. It sets both bits in the solicitation. In the advertisement, if the FCF has a preference, it sets only one bit, basically deciding which scheme to use, otherwise it sets both and leave the decision to the Enode.

On receiving the advertisement messages, the Enode builds a list of the FCFs that have replied with the associated capabilities and priorities ("H2's FCF List" in Figure 51). At this point the Enode can start the next phase of the FIP protocol, i.e. login into the Fabric.

The Enode selects some FCFs from the list, according to its capabilities and configuration policies. For HA (High Availability) it probably selects at least two FCFs and it logs into them.

The login is accomplished with the classic FLOGI message, but this message is carried inside FIP, instead of inside FCoE. The reason for this decision is to make this message easy to intercept by intermediate Ethernet switches. Ethernet switches must monitor all the FIP protocol to open and close ACLs (Ac-

FIP Operation Code = 02h	Reserved	SubCode = 01h
Descriptor List Length = 152	Flags	
•Type = 7	•Len = 144	•Reserved
•FLOGI Request •NPIV FDISC Request •Fabric LOGO Request •(No SOF/EOF / FC-CRC?)		
•Type = 2	•Len = 8	
•MAC Address (VN_Port MAC Address proposed/used by ENode)		

Figure 54: FIP FLOGI/FDISC/LOGO Request

cess Control Lists) in order to protect the Storage traffic from intentional or unintentional (erroneous configuration) attacks.

Figure 54 shows the structure of an FLOGI message encapsulated into FIP. There is an additional advantage of this encapsulation, i.e. the possibility to

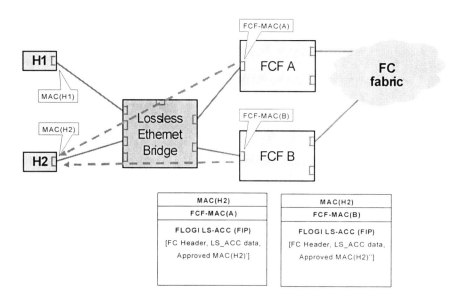

Figure 55: FIP: FLOGI LS_ACC

FIP Operation Code = 02h		Reserved	SubCode = 02h
Descriptor List Length = 152		Flags	
•Type = 7	•Len = 144	•Reserved	
•FLOGI LS_ACC •NPIV FDISC LS_ACC •Fabric LOGO LS_ACC •(No SOF/EOF / FC-CRC?)			
•Type = 2	•Len = 8		
•MAC Address (VN_Port MAC Address approved or assigned by FCF)			

Figure 56: FIP FLOGI/FDISC/LOGO LS_ACC

add an extra field to the FLOGI message, for example the proposed VN_Port MAC address.

If the FCF accepts the FLOGI (see Figure 55), it generates a FLOGI Accept (LS_ACC) that is also carried in FIP to be easily intercepted by the Lossless Ethernet Bridge to complete the ACL setup.

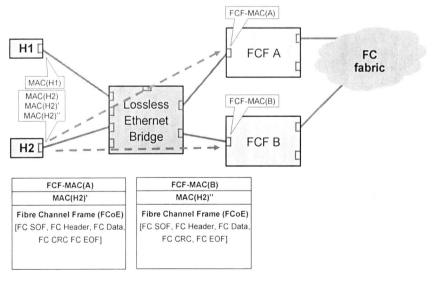

Figure 57: FCoE: Data Transfer

In practice the Lossless Ethernet Bridges redirect the FIP protocol to their supervisor, while they switch in hardware the FCoE frames as regular Ethernet frames.

The format of the LS_ACC is shown in Figure 56.

The LS_ACC concludes the setup process and the FIP protocol: the Enode can start sending regular FC frames using FCoE as shown in Figure 57.

Please note that the notation MAC(H2)' and MAC(H2)" indicates that H2 may use two different FPMAs, one to talk with FCF A and one for FCF B.

3.10. CNAs: Converged Network Adapters

CNA (Converged Network Adapter) is the commercial name for the Enode. It is a PCI Express adapter that contains both the HBA and the NIC functionalities. CNAs are available as regular PCI Express boards and also in mezzanine form factor for the server blades of the blade servers.

CNAs will soon start to appear on the mother board, an arrangement often called LOM (LAN On Motherboard). This will make FC even more mainstream in the Data Center, since it will be available basically "for free" on the motherboard.

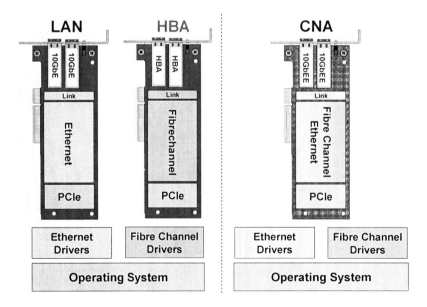

Figure 58: Comparison of NIC, HBA and CNA

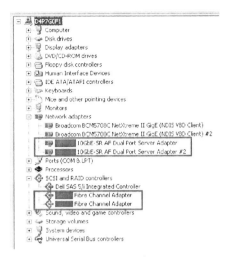

Figure 59: How Windows sees a CNA

The initial CNA vendors are the today HBA vendors. The most common structure of a CNA is illustrated in Figure 58.

The interesting aspect of this architecture is that the view from Operating System is of two separate devices that run standard drivers: a dual port Ethernet adapter and a dual port Fibre Channel HBA.

From the SCSI perspective, the host is not capable of distinguishing that SCSI is running over FCoE instead of over FC.

From a FC perspective, the only additional piece is the FIP protocol that is implemented by the FCoE controller inside the CNA or in the CNA drivers.

The management tools can run unmodified, since they still see FC; over time, they will be extended to be able to monitor also the Ethernet component of FCoE.

Figure 59 shows how Microsoft Windows sees a CNA:

- In the "Network Adapters" section of the Control Panel it sees two Ethernet NIC;
- In the "SCSI and RAID controllers" section of the Control Panel it sees two Fibre Channel Adapters.

Since the CNAs are manufactured by the same companies that manufacture HBAs, they are the first host solution to be certified by the Storage Array vendors.

However, in the medium term also the NIC vendors will offer FCoE capabilities and start to sell CNAs, and this will reshape the overall NIC/HBA/CNA market.

3.11. FCoE: Open Software

FCoE can also be implemented entirely in software. This may be particularly interesting for servers that perform non-I/O intensive operations, but that have the need to access the storage arrays.

A complete software implementation of FCoE exists in the public domain in an open-source project (www.Open-FCoE.org) [13]. This will soon start to appear in Linux distributions like RedHat or SuSE and later possibly also in Microsoft Windows.

The software architecture of this project is shown in Figure 60. On the left side there is the simplified software architecture of a storage stack over a CNA, on the right side the open FCoE software architecture.

Applications not only are unaware of running over FCoE instead of FC, but they are also unaware that FCoE is implemented in software.

This approach clearly loads the CPU more than in the case of a CNA, but is predicated on the fact that, with advent of multi-core CPUs, "CPU cycles are abundant" and some can be spared for FCoE.

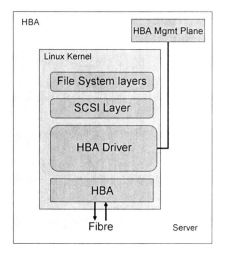 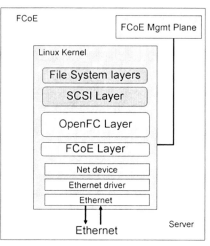

Figure 60: Open-FCoE Software

Figure 61: Wireshark Screen Shot

This approach does not require any particular hardware, just the classical NIC available on the motherboard.

3.12. FCoE: WireShark

Wireshark (http://wireshark.org/) is the most used public domain protocol analyzer, previously known as Ethereal. It captures and displays network traffic and it has been extended to provide a full decode of FCoE.

Figure 61 shows a screenshot of Wireshark while decoding an ELS PRLI FC frame carried over FCoE.

Example traces can be downloaded from:

- http://wiki.wireshark.org/SampleCaptures

3.13. FCoE FAQ

In spending time evangelizing FCoE to the industry and to the customers, I have noticed that there are few questions that periodically pop-up. I will try to answer them in the following sections.

3.13.1. Is FCoE routable?

Allow me to reply with a question: are you sure this is the question you want to ask or is it instead "is FCoE IP routable?"

The answer to the second question is clearly: NO!

It is evident, by looking at Figure 62, that FCoE does not have an IP layer and therefore it is not IP routable. This was not an oversight; it was a conscious design decision. As the famous French writer Antoine de Saint-Exupery (1900 - 1944) used to say: "Perfection is achieved, not when there is nothing more to add, but when there is nothing left to take away."

FCoE is simple and it contains the minimum indispensable to carry FC over Ethernet, and nothing else.

The industry has already developed other standards to carry FC over IP, in particular FCIP (see Figure 62). FCIP is "IP routable" and it is already part of the FC-BB-5 standard that will contain also FCoE.

Figure 63 shows a Disaster recovery solution in which two Data Centers that use FCoE are connected with an IP network using FCIP.

Data that need to be sent to a remote site hops on a FCIP link via a gateway and similarly enter a FCoE network via a FCIP gateway. File Backups to Tape Libraries and Recovery from these devices continue to use existing backup and recovery tools.

Coming back to the original question "Is FCoE routable?" another possible

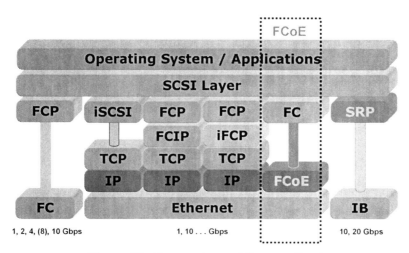

Figure 62: Comparison of Storage Stacks

Figure 63: FCoE and FCIP

answer is: "FCoE, like FC, can be routed by FCoE switches". FCoE switches may forward FC frames across different Ethernet clouds according to the destination FC_ID.

Soon we will start to see on the market multiprotocol routers that implement the FCF functionality in addition to IPv4 and IPv6 routing (similar to the one depicted in Figure 64).

Figure 64 shows that the IPv4 and IPv6 routers operate across VLANs, while each FCF operates inside a VLAN. Multiple FCFs are interconnected by an IFR (Inter Fabric Routing) module.

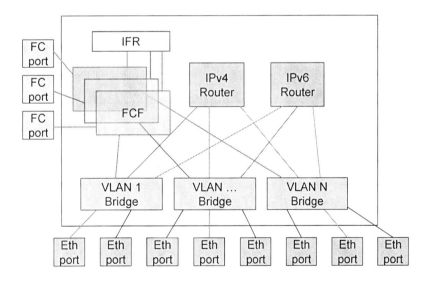

Figure 64: Multiprotocol Router with FCoE capability

3.13.2. Is FCoE the iSCSI killer?

FCoE was not designed to kill iSCSI. There are many applications of iSCSI that will never be covered by FCoE, in particular in the low-end systems, and in the small remote branch offices, where IP connectivity is of paramount importance.

FCoE was designed to be a better fit of iSCSI in the Data Center. Most large Data Centers have a huge install base of Fibre Channel and want a technology that maintains the FC management model. FCoE is such a technology, while iSCSI, even if better, is simply different (there is no FC layer in iSCSI) and therefore not so seamless to integrate into an existing FC environment as FCoE.

FCoE does not require TCP and this is seen as an advantage by many customers. Also, FCoE supports SAN booting as well as FC does and definitively better than iSCSI.

FCoE will take a large share of the SAN market, it will not kill iSCSI, but it will reduce for sure its potential market.

4. Case Studies

This section illustrates possible adoption scenarios of I/O consolidation solutions based on Ethernet and FCoE (see Section 3).

Figure 65 depicts the common architecture of a current Data Center.

The servers at the bottom of the picture have multiple 1 Gigabit Ethernet cards that are typically configured in NIC teaming (aka bonding). These cards connect to Access Layer Ethernet switches like the Catalyst 6509 that can be configured at layer 2 (bridging only) or at layer 3 (IP routing). These Ethernet switches are typically aggregated at a higher level by distribution switches that are normally configured to work at layer 3 (i.e., IP routers). The Distribution Ethernet switches may also host service blades for firewalling, load balancing and intrusion detection. They connect to the LAN core, aka Intranet, and through it possibly to the Internet.

The same servers are equipped with HBAs and connected to Fibre Channel access switches like the MDS 9500. The Fibre Channel topology normally consists of two separates Fabrics (SAN-A and SAN-B) not interconnected. The access layers switches are connected to FC distribution layer switches, aka FC directors.

I/O consolidation is particularly valuable in the connection of servers to the access switches. In fact this allows a strong reduction of interface cards to be installed in the servers, number and types of cables and switch ports.

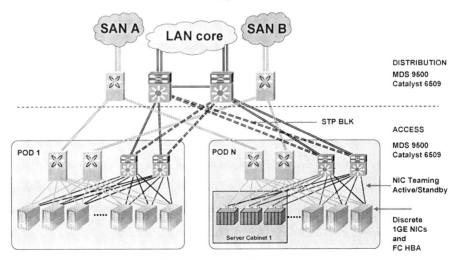

Figure 65: Current Data Center Environment

4.1.   I/O Consolidation with Discrete Servers

Figure 66 shows a first attempt at I/O consolidation in a Data Center that uses discrete rack mounted servers. CNAs are installed on the servers and the Ethernet and FC access level switches have been consolidated into FCoE switches (in the example, the Cisco Nexus 5000). The I/O from the server to the FCoE switches is consolidated over 10GE, for example, using copper twinax cables.

The FCoE switches operate at layer 2, since this greatly simplifies the movement of Virtual Machine in the Data Center (the MAC and IP addresses can be easily moved inside layer 2 domains).

The distribution layer is unchanged. The FC uplinks from the FCoE switches plug into the FC SAN-A and SAN-B, which are two separate SANs. Two separate instances of FSPF run one on SAN-A and the other on SAN-B. Each instance manages its topology, and all the FC links are active and forward traffic.

The situation is different on the Ethernet side, where there is a single LAN backbone.

By default the Spanning Tree Protocol (STP) runs on the Ethernet meshes present between the access and the distribution layer switches and prune them to a tree, by blocking Ethernet ports. Links connected to blocked ports don't

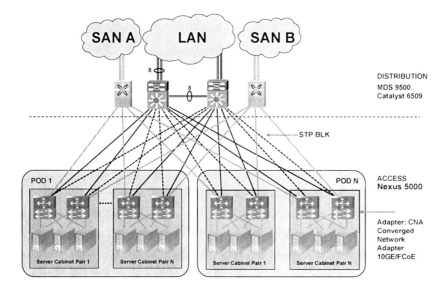

Figure 66: I/O Consolidation with Spanning Tree

carry any traffic. This can be mitigated by running STP per VLAN and by blocking different links on different VLANs, thus load balancing the traffic.

This can be avoided in three different ways: by using a feature called "Ethernet Host Virtualizer" on the access switches, by using a feature called VSS (Virtual Switch) on the distribution switches, or by running Layer 2 multipath between the access switches and the distribution switches.

Figure 67 shows the Ethernet Host Virtualizer (see Section 2.8.5). When configured in this mode, the access switches don't forward the traffic received from the distribution switches back to the distribution switches: basically, they do not create meshes. The hosts connected to the access switches are statically load-balanced among the Ethernet uplinks. The resulting effect is that there is no need to run STP between the access layer and the distribution layer. STP still runs to protect from erroneous configurations, but it does not see any mesh and therefore it does not prune any link. All the links remain in forwarding state.

Implementations pay particular attention to how the multicast and broadcast traffic is forwarded to avoid creating broadcast storms and duplicating broadcast/multicast frames.

Figure 68 shows an alternative way to skin the cat. In this configuration the distribution switches are configured in VSS (Virtual Switch) mode and therefore

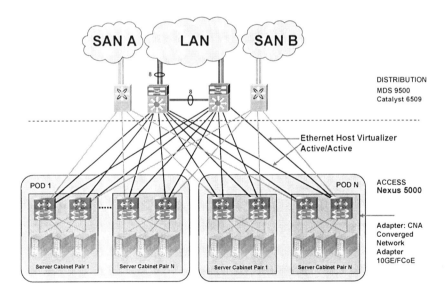

Figure 67: I/O Consolidation with Ethernet Host Virtualizer

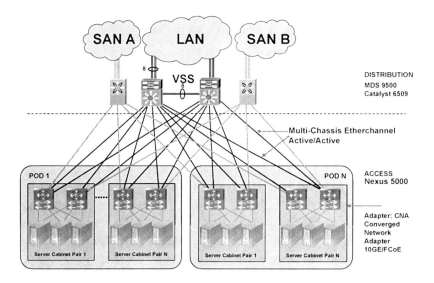

Figure 68: I/O Consolidation with VSS and Etherchannel

they present themselves to the outside world as a single switch. This allows the access switches to use Etherchannel (aka Link Aggregation), since they cannot tell that they are speaking to two separate boxes. This particular configuration of Etherchannel is also called Multi-Chassis Etherchannel.

Etherchannel load balances the traffic on all the links that are aggregated by using some hash function that normally considers fields in the layer 2 and layer 3 headers.

Again, the STP does not see any mesh, since it considers each Etherchannel a single link and therefore does not prune any link.

4.2.   Top-of-Rack Consolidated I/O

The three previous solutions lead to an implementation that uses top-of-rack switches. The reason is that the twinax cable presents a strong economical convenience, it rationalizes nicely and simplifies the server cabling, but it is limited to 10 meters (i.e., 33 feet).

With this distance limitation it is possible to connect servers located in one to five racks. The number of servers per rack is a function of the power and cooling infrastructure of the Data Center. Typical numbers go from 10 servers (approx. 7.5KW) to 20 servers (approx. 15kW) per rack.

For example, the Cisco Nexus 5000 has a configuration dedicated to intercon-

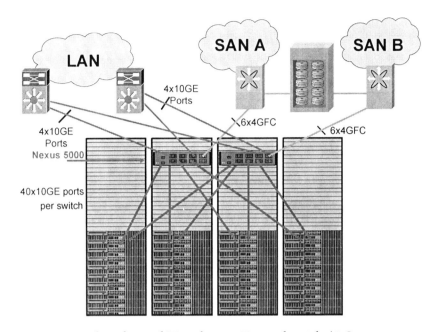

Figure 69: Physical Topology – Example with 40 Servers

nect 40 servers (not all cables are shown). Figure 69 shows a configuration with 10 servers per rack that uses four racks. A similar configuration can be obtained in two racks with 20 servers per rack.

Two Nexus 5000 are installed as top-of-racks and the servers are equipped with dual ports CNAs. One port is connected to one Nexus 5000 and the other to the other Nexus 5000.

The Nexus 5000 are then connected to a common LAN for the native Ethernet traffic, but to two separate SANs for the FC traffic.

Figure 69: The Distribution switches can be placed at the end of the row and, even if the distance is superior to 10 meters, and the connection therefore requires fiber optics, the cost of the optics is subdivided on a pool of servers and therefore it is not so critical.

Let's try to estimate the cabling saving of this solution. Figure 70 contains an example in which it is possible to see that the number of adapters is clearly divided by 2 and the number of cables is reduced from 72 to 40 (it is not exactly half, since the cables that connect the distribution switches are not consolidated).

Figure 71 shows that to build a large pool of servers it is sufficient to repeat the

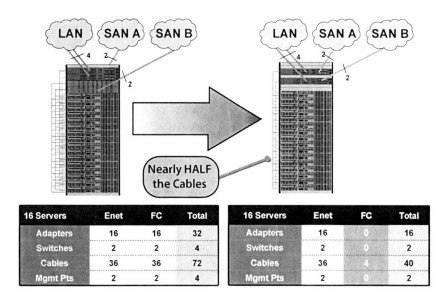

Figure 70: Cable Reduction with FCoE and 10GE

basic building block numerous times.

Assuming distributions switches with 128 ports total, of which 100 ports are dedicated to connect access switches, it is possible to build a server pool with 1,000 servers. With larger port count in the distribution layer switches, or a larger number of distribution switches, it is possible to build even larger server pools.

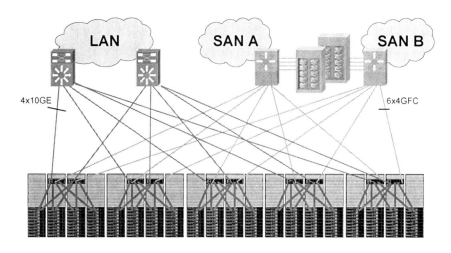

Figure 71: Physical Topology – Example with 200 servers

4.3. Example with Blade Servers

What explained in the previous examples applies to blade servers as well. A blade server is an enclosure that provides power, cooling and interconnectivity to a set of server blades.

Blade servers may host switch blades capable of switching Ethernet, Fibre Channel or Infiniband.

Figure 72 shows an example of blade servers in which no switch blades are installed, but the connectivity is provided through copper pass-through modules. These are passive cards that present one twinax connection on the outside for each server blade in the blade server.

In this configuration nothing really changes compared to the previous examples: the servers have a different form factor, nothing else.

Figure 73 shows blade servers equipped with two 10 GE Ethernet switches.

Compared to Figure 72 there are the following differences:

- The number of cables between the blade servers and the access switches is reduced, since the Ethernet switches perform statistical

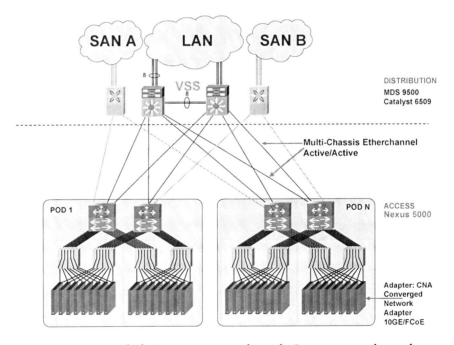

Figure 72: Blade Servers: example with Copper pass-through

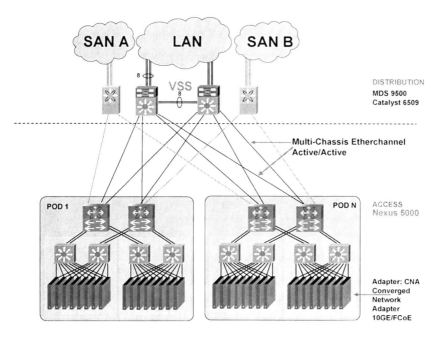

Figure 73: Blade Server: example with pure Ethernet Switch

multiplexing of traffic, for example by a factor of two or four;

- The number of access switches is reduced by the same factor, since less ports are needed at the access layer;

- Local switching among server blades can be enabled, even if sometimes it is problematic to enforce consistent policies when part of the switching happens in the blade servers;

- The cost of the Ethernet switch in the blade server is higher than the cost of the pass-through module;

- The Ethernet switch in the blade server is an additional element that requires management.

Figure 74 shows a rack with two Nexus 5000 and two blade servers, each with 10 server blades and two Ethernet switch blades. Each blade server uses one 10 GE uplink from each Ethernet switch blade to each Nexus 5000. This is an extreme example of traffic multiplexing but, if more bandwidth is desired, additional links can be inserted in parallel and configured in Etherchannel.

Figure 75 shows 5 racks, each one with two blade servers connected to a pair of Nexus 5000 using twinax cabling.

Again this can be repeated multiple times to create larger server pools.

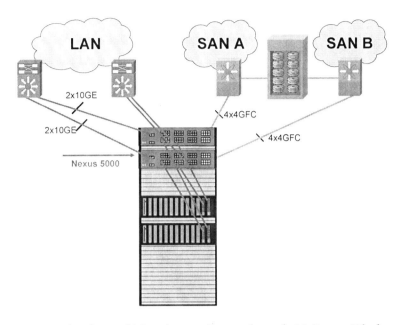

Figure 74: Physical Topology – Example with 20 Server Blades

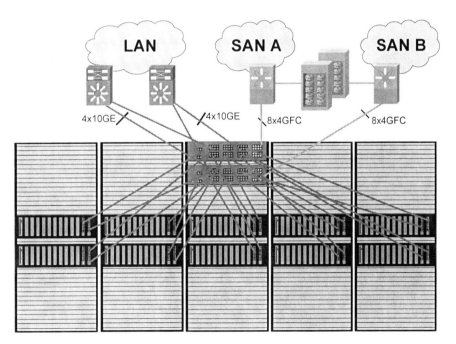

Figure 75: Physical Topology – Example with 100 Server Blades

4.4.   Updating the distribution layer

Figure 76 shows a possible evolution of the distribution layer limited to the Ethernet portion.

The Catalyst 6509 can be replaced by the larger and more powerful Nexus 7000, which also supports layer 2 multipath. The STP protocol can be eliminated between the Access and the Distribution layer, greatly increasing and rationalizing the bandwidth available.

Figure 77 shows the last step: the Nexus 7000 also implements the FCF functionality and therefore the MDS switches are no longer needed. FC cards are installed on the Nexus 7000 and they provide connectivity to the FC SANs.

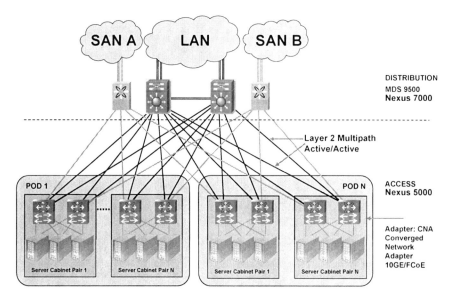

Figure 76: Ethernet Layer 2 Multipath

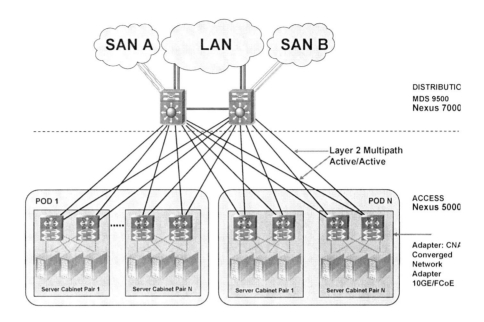

Figure 77: Consolidation in the Distribution Layer

5.   Bibliography

5.1.   PCI Express

[1]   http://en.wikipedia.org/wiki/Pci_express

5.2.   IEEE 802.3

[2]   http://standards.ieee.org/getieee802/802.3.html

5.3.   Improvements to Ethernet

[3]   http://www.nuovasystems.com/EthernetEnhancements-Final.pdf

5.4.   IEEE 802.1 activities

[4]   http://www.ieee802.org/1/files/public/docs2007/new-cm-barrass-pause-proposal.pdf

[5]   http://www.ieee802.org/1/files/public/docs2007/new-cm-pelissier-enabling-block-storage-0705-v01.pdf

[6]   http://www.ieee802.org/1/files/public/docs2007/au-ko-fabric-convergence-0507.pdf

[7]   http://www.ieee802.org/1/pages/802.1au.html

[8]   http://www.ieee802.org/1/files/public/docs2008/az-wadekar-dcbcxp-overview-rev0.2.pdf

[9]   http://www.ieee802.org/1/files/public/docs2007/new-wadekar-priority-groups-1107-v1.pdf

[10]   http://www.ieee802.org/1/pages/802.1az.html

5.5.   FCoE

[11]   http://www.fcoe.com/

[12]   http://www.t11.org/

[13]   http://www.open-fcoe.org/

[14]   http://www.fibrechannel.org/OVERVIEW/FCIA_SNW_FCoE_WP_Final.pdf

5.6.   TRILL

[15]   http://www.ietf.org/html.charters/trill-charter.html

5.7.   Virtualization

[16] http://www.vmware.com/support/esx21/doc/esx21admin_MACad-
     dress.html

6. Glossary, Tables and Index

6.1. Glossary

- 10GBASE-T: A standard for 10 Gigabit Ethernet over twisted pair

- 10GE: 10 Gigabit Ethernet, see also IEEE 802.3

- 802.1: An IEEE standard for LAN Bridging & Management

- 802.1Q: An IEEE standard for bridges, VLANs, STP, priorities

- 802.3: The Ethernet standard

- ACL: Access Control List, a filtering mechanism implemented by switches and routers

- Aka: Also Known As

- AQM: Active Queue Management, a traffic management technique

- B2B: Buffer-to-Buffer, as in Buffer-to-Buffer credits for FC, a technique to not lose frames

- BCN: Backward Congestion Notification, a congestion management algorithm

- CEE: Converged Enhanced Ethernet, a term used to indicate an evolution of Ethernet

- CNA: Converged Network Adapter, the name of a unified host adapter that support both LAN and SAN traffic

- CRC: Cyclic Redundancy Check is a function used to verify frame integrity

- DCB: Data Center Bridging, a set of IEEE standardization activities

- DCBX: Data Center Bridging eXchange, a configuration protocol

- DCE: Data Center Ethernet, a term used to indicate an evolution of Ethernet

- dNS: the Fibre Channel domain Name Server

- DWRR: Deficit Weighted Round Robin, a scheduling algorithm to achieve bandwidth management

- Enode: a host or a storage array in Fibre Channel parlance
- F_Port: A Fibre Channel port that connects to Enodes
- FC: Fibre Channel
- FC_ID: Fibre Channel address, more properly N_Port_ID
- FC-BB-5: the working group of T11 that is standardizing FCoE
- FCC: Fibre Channel Congestion Control, a Cisco technique
- FCF: Fibre Channel Forwarder, a component of an FCoE switch
- FCIP: Fibre Channel over IP, a standard to carry FC over an IP network
- FCoE: Fibre Channel over Ethernet
- FCS: Frame Check Sequence is a function used to verify frame integrity
- FIP: FCoE Initialization Protocol
- FLOGI: Fabric Login
- FPMA: Fabric Provide MAC Address
- FSPF: Fibre Channel Shortest Path First
- HBA: Host Bus Adapter, the adapter that implement the Enode functionality in the host and in the storage array
- HCA: Host Channel Adapter, the IB adapter in the host
- HDLC: High-Level Data Link Control, a serial protocol
- HOL blocking: Head OF Line blocking is a negative effect that may cause congestion spreading
- HPC: High Performance Computing
- IB: Infiniband, a standard network for HPC
- IEEE: Institute of Electrical and Electronics Engineers (www.ieee.org)
- IETF: Internet Engineering Task Force (www.ietf.org)
- IP: Internet Protocol
- IPC: Inter Process Communication
- IPv4: Internet Protocol version 4
- IPv6: Internet Protocol version 6

- iSCSI: Internet SCSI, i.e. SCSI over TCP

- ISO: The International Organization for Standardization is an international standard-setting body composed of representatives from various national standards organizations.

- LAN: Local Area Network

- LAPB: Link Access Protocol, Balanced, a serial protocol

- Layer 2: Layer 2 of the ISO model, also called datalink. In the Data Center the dominant layer 2 is Ethernet.

- Layer 3: Layer 3 of the ISO model, also called internetworking. The dominant Layer 3 is IP, both IPv4 and IPv6

- Layer 4: Layer 4 of the ISO model, also called transport. The dominant Layer 4 is TCP

- Layer 7: Layer 7 of the ISO model, also called application. It contains all applications that use the network

- LLC: Logical Link Control, a key protocol in IEEE 802.1

- LLC2: LLC used with reliable delivery

- LLDP: Link Layer Discovery Protocol, an Ethernet configuration protocol, aka IEEE 802.1AB

- MPI: Message Passing Interface, an IPC API

- N_Port_ID: Fibre Channel address, aka FC_ID

- NFS: Network File System

- NIC: Network Interface Card

- PCI: Peripheral Component Interconnect, a standard I/O bus

- PFC: Priority Flow Control, aka PPP

- PPP: Per Priority Pause, aka PFC

- QCN: Quantized Congestion Notification, a congestion management algorithm

- R_RDY: Receiver Ready is an ordered set used in Fibre Channel to replenish buffer-to-buffer credits

- RDMA: Remote Direct Memory Access, an IPC technque

- RDS: Reliable Datagram Service, an IPC interface used by databases

- RED: Random Early Detect, an AQM technique

- RSCN: Registered State Change Notification, an event notification protocol in Fibre Channel

- SAN: Storage Area Network

- SCSI: Small Computer System Interface

- SDP: Socket Direct Protocol, an IPC interface that mimics TCP sockets

- SFP+: Small Form-Factor Pluggable transceiver

- SPMA: Server Provide MAC Address

- SPT: Spanning Tree Protocol, see also IEEE 802.1Q

- T11: Technical Committee 11, which is the committee responsible for Fibre Channel (www.t11.org)

- TCP: Transmission Control Protocol, a transport layer protocol in IP

- TRILL: Transparent Interconnection of Lots of Links Working Group within the IETF

- Twinax: a twin micro-coaxial copper cable used for 10GE

- VE_Port: an FCoE port on an FCoE switch used to interconnect another FCoE switch

- VF_Port: an FCoE port on an FCoE switch used to interconnect an Enode

- VLAN: Virtual LAN

- VN_Port: an FCoE port on an eNODE used to interconnect to an FCoE switch

- Wireshark: (http://www.wireshark.org/) a public domain protocol analyzer

6.2. Figures

6.3.   Tables

6.4.   Index

Printed in the United States
208894BV00003B/123-128/P